A Practical Guide to Bio-inspired Design

Helena Hashemi Farzaneh · Udo Lindemann

A Practical Guide to Bio-inspired Design

Springer Vieweg

Helena Hashemi Farzaneh
Laboratory for Product Development
and Lightweight Design
Technical University of Munich
Garching, Germany

Udo Lindemann
Laboratory for Product Development
and Lightweight Design
Technical University of Munich
Garching, Germany

ISBN 978-3-662-57683-0 ISBN 978-3-662-57684-7 (eBook)
https://doi.org/10.1007/978-3-662-57684-7

Library of Congress Control Number: 2018950041

Springer Vieweg

Printed on acid-free paper

This Springer Vieweg imprint is published by the registered company Springer-Verlag GmbH, DE part of Springer Nature
The registered company address is: Heidelberger Platz 3, 14197 Berlin, Germany

Preface

Bio-inspired design – mentioning this term to anyone immediately draws attention. Most people feel intrigued by the idea of bio-inspired design; looking at nature and its solutions to solve technical challenges. Nature is fascinating for us. Since the beginnings of humanity it has inspired our culture. Arts, literature, architecture, design, and technology have all profited from nature as a source of inspiration. Therefore, bio-inspired design seems to present obvious solutions to us, designers from all disciplines, whenever we are stuck and have no ideas on how to solve a technical challenge.

Still, bio-inspiration is not a very widespread innovation method. For example, in Germany, only very few universities offer courses in bio-inspired design. Even the number of lectures addressing bio-inspired design is quite small. If we ask professional engineers and designers, the use of bio-inspired design methods seems to be almost unknown in research and development departments of engineering companies. If bio-inspired design is applied, it often does not go beyond regarding obvious biological systems as inspirations, for example analyzing humans for developing a humanoid robot. The application of methodological bio-inspired design seems to be rare. There are just a few cases in which engineering designers have systematically searched for the most suitable biological inspiration for their specific technical challenge. The cases in which biologists have attempted to transfer a biological solution over to a technical application are even fewer.

Why is this so? Almost all of us know some classic examples of bio-inspired design. Leonardo da Vinci's drawings of technical machines based on natural inspirations are famous. Very well-known is also the invention of Velcro® in the mid-20th century: the Swiss engineer Georges de Mestral was on a walk with his dog when the burrs clinging to his dog's fur caught his attention. He analyzed these burrs under the microscope and developed the famous hook-and-loop fastener. In the mid-20th century the research interest in bio-inspired design also increased. Many authors name the *Bionics Symposium* that took place at the Wright-Patterson Air Force Base, Ohio in 1960 as the beginning of systematic research efforts towards bio-inspired design. The Symposium was entitled "*Living Prototypes – the Key to New Technology*" and predicted that technical fields such as electronics, aeronautics, and guidance and control could profit from bio-inspired design. A major challenge of bio-inspired design was attributed to the complexity of biological systems (Keto 1960).

The complexity of biological systems is certainly one of the major challenges of bio-inspired design, even today. Despite the enormous advances of the discipline of biology, biological knowledge is difficult to access for laypersons and difficult to apply in bio-inspired designs. Providing support for this challenge is the main aim of this book. We believe that the use of outlined methods can greatly facilitate bio-inspired design and improve the feasibility of its outcomes. Bio-inspired design methods are therefore the key to spreading bio-inspired design as an innovation method.

The entire team of the Institute of Product Development at the Technical University of Munich supported us in conceptualizing this book. We would like to thank the whole team for the valuable discussions about bio-inspired design and bio-inspired design methods. In particular, we would like to thank Maria Katharina Helms for her contribution to bio-inspired design research at the Institute and to the concept of this book. We would also like to thank Eva Körner for developing the visual concept of the fact sheets.

Special thanks go to the undergraduate students who supported us: Marie-Luise Domke, who conducted the interviews for the best practice examples and supported the content-related investigation and formatting, Anton Doqi, who designed the layout of the fact sheets as well as many of the figures and sketches*, and Jonathan Creed and Daniel Ledda for proofreading the book.

This book project would not have been possible without the financial support of the Technical University of Munich. Therefore, we would like to thank the President, Prof. Dr. Dr. h.c. mult. Wolfgang A. Herrmann and all members of the TUM Leonardo Da Vinci Centre for sponsoring our work.

For the excellent collaboration and support, we would also like to thank the publishing house, particularly Thomas Lehnert, Ulrike Butz and Peter Sollfrank.

With this book, we hope to contribute to the systematic and methodological application of bio-inspired design in teaching and practice. The potential of bio-inspired design for technical innovation is enormous and still largely unexplored. We hope that this book supports engineers, biologists, and interdisciplinary collaborations in changing this.

Munich, May 2018,

Udo Lindemann and Helena Hashemi Farzaneh

**Figures or Sketches designed by Anton Doqi:*
Figure 1-4, Figure 1-5, Figure 2-1, Figure 2-2, Figure 2-3, Figure 2-7, Figure 2-10, Figure 2-11, Figure 2-12, Figure 3-2, Figure 3-4, Figure 3-5, Figure 3-6, Figure 3-7, Figure 3-8, Figure 3-9, Figure 3-10, Figure 4-7, Figure 4-8, Figure 4-9, Figure 4-10, Figure 4-11, Figure 4-13, Figure 5-8, Figure 6-6

Moreover, Anton designed all Figures and Sketches on the fact sheets.

Contents

Best Practice Examples

1 Introduction

Bio-inspired design is a transdisciplinary approach; it links biology and technology, generating knowledge beyond the disciplinary borders. Helms et al. (2009) defined biologically inspired design as the approach to using "*analogies to biological systems to develop solutions for engineering problems*." In this book, we understand bio-inspired design in a slightly broader sense: *Bio-inspired design is the application of knowledge of biological systems in research and development for technical inventions and innovations.*

Bio-inspired design is a relatively new term. More wide-spread terms are biomimetics (e.g. Lepora et al. 2013) and biomimicry (e.g. Benyus 1997).

Bio-inspired design has been used to develop innovative technical solutions in almost every engineering field. When the idea of analyzing biological systems systematically emerged in the mid-20th century as *bionics*, the main focus was on control and communication systems (Keto 1960). Keto (1960) proposed applying bio-inspired control and communication in electronics, aeronautics, and navigation. However, historic examples of bio-inspired design already include further application areas. For example, the aerodynamics of biological systems were frequently regarded as models to conceptualize parachutes or early flying machines (Nachtigall and Wisser 2015, p.6). Modern examples of bio-inspired design are applied in technical fields such as mechanical engineering, architecture, industrial design, software engineering, robotics, medical technology, and so on. In this book, we show ten *Best practice examples* to illustrate the different approaches and application areas of bio-inspired design.

Despite its high potential for innovation, the application of bio-inspired design presents challenges. Major challenges are disciplinary barriers, i.e. the goals, working approaches, and the fact that language of biology and technical product development are fundamentally different. Biology is a natural science; it aims to understand living systems. Technical product development, or engineering design, aims to develop technical systems. The difference between the biological goal of *understanding* and the technical goal of *development* reflects the different working approaches of biologists and engineers. This clash frequently causes problems in bio-inspired design.

For example, technical systems should fulfil defined functions. For example, a car must transport people, or a coffee machine must brew coffee. Engineers have to develop technical systems which fulfil these functions, usually in a robust and simple manner to reduce faults and costs. If engineers look at nature and try to mimic biological functions, this may prove difficult. Firstly, biological systems are multifunctional, and many of their functions are unknown. Secondly, they often fulfil their functions by means of multiple solutions on several scales. Biologists, on the other hand, analyze these solutions in detail. However, it is very rare that they dare to simplify their findings to such an extent that a comparably "simple" technical solution can be deduced.

H. Hashemi Farzaneh and U. Lindemann, *A Practical Guide to Bio-inspired Design*, https://doi.org/10.1007/978-3-662-57684-7_1

The aim of this book is therefore to support biologists and engineers (as well as everyone else) to apply bio-inspired design in practice. We want to bridge the gap between biology and technical product development by divulging a selection of methods that have been developed in bio-inspired design research. We have structured this book according to five bio-inspired design-specific activities, namely: planning, abstraction, search, analysis and comparison, and transfer. For each activity, we have selected methods that support the systematic application of bio-inspired design. We believe that the purposeful, reflective use of these methods is the key to improving the outcomes of bio-inspired design projects. *Systematic use of methods* does not imply rigidly following a method even though the outcomes are not convincing. Instead, designers, meaning both biologists and engineers, should always reflect on the usefulness of a method in their context and apply methods flexibly.

We will take into account the variety of bio-inspired design projects by indicating the suitability of bio-inspired design methods for the two distinct bio-inspired design approaches (biology push and technology pull) and three scenarios which range from quick application of bio-inspired design to long-term projects.

To illustrate the use of methods we will use one guiding biological example and three hypothetical technical applications. As mentioned above, we will furthermore present ten *Best practice examples* to showcase different possibilities of applying bio-inspired design.

In the following sub-sections, we will explain more details regarding:

- Why should you apply bio-inspired design and use this book?
  Ten aspects highlight why bio-inspired design has such a high potential for advancing technical product development and biology (1.1).
- How can you use this book?
  We explain our concepts of bio-inspired design activities, approaches and scenarios (1.2).
- What about our guiding example?
  We explain our guiding biological example, the tree and torrent frogs' toe pads, and three hypothetical technical application ideas (1.3)

1.1 Why this Book? – 10 Amazing Aspects of Bio-inspired Design

Abundance of Biological Solutions

It has often been said already, but it is true: biology offers a vast pool of possible solutions to technical problems. In 2006, 2 million known species were counted – and this pool is still growing. Every year, approximately 12,000 new animal, plant, and microorganisms are described in scientific publications (Streit 2006). In the digital age, the access to biological inspirations is also becoming easier. Digital databases such as asknature.org provide access to several thousands of biological inspirations (Deldin and Schuknecht 2014).

Optimization

"Biological systems have been optimized by evolution" – this is a common saying that supports bio-inspired design (Nachtigall 2002, p.357–388; Fish and Beneski 2014). We must be careful in assuming that, if a biological system is optimized, the bio-inspired design is optimized as well. Biological systems must fulfil multiple goals and we do not know all of them. Still, biological systems often fulfil their goals in a particularly resource-efficient way. This characteristic is very interesting for bio-inspired design. Moreover, despite having different ancestors, numerous biological species have developed similar solutions to similar problems, that is, they *converge*. For example, African succulents and American cactuses strongly resemble each other with their water-retaining, fleshy shapes – an adaptation to hot, dry climate. This convergence through adaptation can be considered a strong indication for an optimized solution (Vogel and Ferrari 2013, p.525–527).

In addition, evolution itself has been studied as inspiration for optimization problems. Numerous evolutionary algorithms have been developed; some use for example evolutionary strategies (ES), such as mutation or recombination for optimization problems (Rechenberg 1972; Nachtigall 2002, p.357–388).

Synergy Effects and Transdisciplinarity

Not only can technical product development profit from bio-inspired design, but biological research as well. Nachtigall (2002) has defined technical biology as the *understanding of nature using technical support*. He argues that technical biology is a prerequisite for bio-inspired design. But how should a biological researcher analyze the technically applicable aspects of a biological system without technical support? The guiding example of this book, the toe pad structure of tree or torrent frogs is a good one. The adhesion capacities of living tree and torrent frogs have been measured by placing them on slowly rotating glass plates (Endlein et al. 2013b). However, explaining the adhesion mechanisms is difficult, because the toe pads' structure is hierarchically built up on several scales. Using real frogs, it is impossible to consider the toe pads' micro- and nanostructures separately. At this point, technical reproductions are needed. Collaborating material researchers and biologists have successfully produced polymer microstructures similar to the frog and examined the influence of different forces under dry and wet conditions (Drotlef et al. 2013). This interdisciplinary collaboration and knowledge transfer resulted in the generation of new transdisciplinary knowledge. In other words, it advances biological research and enables future bio-inspired designs.

Multifunctionality of Biological Systems

Biological systems have to fulfil multiple functions using limited resources. In particular, if we have a biological system in mind, we will be inspired to expand the functionality of a technical application. These so-called biology push projects tend to result in multifunctional bio-inspired products (Helms et al. 2009). For example, many biological materials, such as bones or shells, resist forces, are lightweight-designed, and self-healing. This combination of functions is certainly interesting for technical materials. Interestingly, a self-healing concrete has already been developed. This concrete is repaired by limestone-producing bacteria which fill cracks (Jonkers 2007).

Integration of Material, Shape, and Function

Form follows function is a design principle often cited in industrial design and architecture. In nature, this principle can be observed frequently – and on multiple levels. Technical systems typically fulfil functions on a few or just one level of scale and are built from monolithic material such as steel. Biological systems often realize functions on multiple scales using both composite material properties and shapes. A good example is the guiding example of this book, the tree or torrent frog's toe pad structure. The toe pad structure's adhesive properties (visible on macroscale) seem to be due to its shape on the micro-level (hexamers) and on the nano-level (pillars). Another example is the iridescent colors of butterflies or beetles. They are not generated by pigments but by interference, diffraction, and scattering of photonic structures on the nano-level (Vukusic and Sambles 2003).

Discovery Beyond that Which is Human-like

It is surprising how much we focus on bio-inspirations similar to us, the human species. Analyzing research publications from two journals of bio-inspired design, we have found that only five percent of the publications use plants as inspiration for bio-inspired designs (Domke and Hashemi Farzaneh 2018). Doing so, we miss out on many biological solutions with high potential for technical application. Biological researchers have, for example, recently discovered that *Boquila trifoliolata*, a climbing plant, can adapt the shape and color of its leaves to the leaves of several host or neighboring plants. This also works if there is no direct contact to the neighboring plants (Gianoli and Carrasco-Urra 2014).

It is still unclear as to how the plant is able to recognize the shape of neighboring leaves. For example, Baluška and Mancuso (2016) suppose that the plant is able to visually perceive shapes and colors.

In any case, *Boquila trifoliolata* must possess some means of recognition, information processing, and control over the color and shape of leaves. This is a surprising capability for an organism that does not have a brain or a similar central controlling organ, but is built up from flexible elements (roots, stem, leaves). It could serve as an inspiration for bio-inspired design of, for example, swarm robots.

Flexibility of Bio-inspired Design

Bio-inspired design is a very flexible design approach. We can quickly search for inspiration through internet searches or creativity workshops. To increase our level of understanding, we can look for existing technical and biological information in research publications. If additional research is required to realize a bio-inspired design, interdisciplinary collaborative research projects are an option.

Moreover, bio-inspired design is not restricted to a particular area of application – a single biological system such as spider silk (see best practice example, p. 140) can inspire applications in technical areas as diverse as medical technology (implant coating), textile engineering (high performance fibers), and home appliances (filters vacuum cleaners).

Different bio-inspired design methods address all these bio-inspired scenarios and application areas. With methodological training, we are able to conduct entirely different development projects.

Sustainability (or not?)

Bio-inspired design is frequently associated with environmentally-friendly, sustainable solutions. There are numerous examples for bio-inspired sustainable products from diverse fields, such as tribology (Nosonovsky and Bhushan 2012) and architecture (John et al. 2005). We must emphasize that bio-inspired designs are not sustainable per se. Even single biological systems on levels beneath the ecosystem level are not sustainable. For example, in Middle European forests, deer often live without natural predators, such as wolves or bears. Consequently, the deer population can grow disproportionately and can seriously damage the forest fauna.

Nevertheless, biology can provide a fundamentally different strategy to sustainable design. The conventional strategy to reach sustainability is to reduce the environmental impact of an existing technical system by, for example, reducing weight. The bio-inspired strategy is not a "reduction" strategy, but it can be based on an analysis of an entire ecosystem, its components, and balancing relations. This view allows conceptualizing bio-inspired technical systems that are sustainable from scratch.

Biodiversity

In addition to its technical and research value, bio-inspired design can have a positive societal impact. It generates attention for the magnificent solutions present in nature and the value of biodiversity. Only if we protect this biodiversity, will we be able to discover new bio-inspirations that can support technological advances. The destruction of nature and the disappearance of biological species also makes nature's blueprints disappear (Benyus 1997, p.292–294; Nader and Hill 1999, p.14). In an age in which industrial development and the protection of nature are still regarded as incompatible opposites, this awareness is crucial.

Biologization

"The 21st century is the century of biology" – this quotation can be taken from several genetic scientists and molecular biologists (Venter and Cohen 2004; Silver 2011; Venter and Cohen 2014). Both research areas have evolved enormously during the last decades. They are the key to understanding the reasons behind biological phenomena we can observe, such as heredity, diseases, behavior and so on. Understanding biology on the molecular level enables us to engineer biological

systems. Moreover, it entails a discipline-spanning approach: the borders between the traditional natural sciences of biology, chemistry, and physics on the one hand, and the technical, engineering disciplines on the other, get blurred. Biological solutions cannot only be imitated by technical means, but biological systems can increasingly be integrated into technical systems. This opens up new possibilities for bio-inspired design and might help to tackle some of the biggest challenges of our century, such as energy consumption and a sustainable economy (Silver 2011).

1.2 How to Use this Book

The main motivation for writing this book is to provide practical support for bio-inspired design projects. Throughout our research in bio-inspired design we have come across numerous publications on bio-inspired design, including a few books. Many of them present very interesting examples of bio-inspired design or categorize different bio-inspired design applications (e.g. bio-inspired robotics, bio-inspired materials etc.). Others describe specific methods from their own research. With this book, we want to go a step beyond: we have collected and selected 27 methods for different bio-inspired design activities and scenarios. Some of the methods were developed decades ago and have been tested in practice. Other methods are rather new and have only been used for teaching bio-inspired design to students. Our main motivation for selecting specific methods was to offer a portfolio of methods to our readers that covers a wide range of bio-inspired design activities and scenarios. This approach entails translating a number of methods to English which have not been published in English before.

Each method supports a specific activity of bio-inspired design, namely: planning, abstraction, search, analysis and comparison, and transfer. We have divided this book into five chapters according to these activities. In each chapter, we explain four to seven methods. Moreover, at the end of the book, a factsheet for each method summarizes the method, its advantages and challenges, and its suitability for different bio-inspired design approaches and scenarios.

How should you, as a practicing biologist or engineer read this book? You can of course read it from beginning to end. However, this might not be the quickest way to get the information that you need. One of our aims when conceptualizing the book was to make the different parts of the book independent – so you can:

- Read the chapter on the activities relevant for your bio-inspired design project
- Use the fact sheets to compare and select suitable methods for your bio-inspired design project
- Read the book from beginning to end for an overview on the whole bio-inspired design process

In the following subsections, we briefly explain what we mean by bio-inspired design activities (1.2.1), approaches (1.2.2) and scenarios (1.2.3).

1.2.1 Bio-inspired Design Activities

Bio-inspired design methods support different activities. To structure this book, we have summarized them under five different headlines:

Planning
Planning a bio-inspired design project differs from planning other technical product development projects or planning biological research projects. The interdisciplinary nature of bio-inspired design poses a challenge for the design procedure, and there can be different starting points for the project (see next section). Moreover, if biologists and engineers collaborate, the integration of their work has to be taken into account.

Abstraction
Abstraction is an important part of bio-inspired design. In fact, designers have to conduct abstraction over and over again throughout a bio-inspired design project: they must abstract a technical task or enable searching for biological inspiration. They must abstract technical and biological systems to make them comparable. They must abstract again to formulate analogies for transfer… and so on.

Search
Search, in particular the search for biological inspiration is addressed by numerous methods for bio-inspired design. It is a key activity when a bio-inspired design project is started from a purely technical viewpoint (technical task, looking for a biological inspiration) or from a purely biological viewpoint (interesting biological solution, looking for a technical application).

Analysis and Comparison
Designers have to analyze and compare both technical and biological systems carefully to apply bio-inspired design successfully. There are numerous examples where bio-inspired designs failed at the prototype stage, because the biological solution could not be applied in a technical environment. An example from the authors' experience is the attempt of a student's project to mimic the flagella movement of bacteria on a much larger scale. Due to the different influence of fluid forces, the rotational flagella movement could not be used to effectively move liquids.

Transfer
The transfer of bio-inspired analogies is the core activity of bio-inspired design. All bio-inspired design projects aim at transferring more or less abstract analogies to develop a technical solution. Researchers have reported a number of challenges of analogical transfer – without sufficient preparation and support designers can fail to transfer suitable analogies.

That said, bio-inspired design projects include of course many other activities. However, these activities are not particular to bio-inspired design, as they are also present in other technical product development projects or in biological research projects. Especially in the early and later phases of a bio-inspired design project, other activities frame the bio-inspired design activities. This is illustrated in Figure 1-1. In the early phases of a bio-inspired design project, biological research can be necessary to deepen the knowledge of the biological system. Moreover, product ideas must be developed – this necessitates research on stakeholders, market, and so on. In the later phases of bio-inspired design projects, the developed bio-inspired solutions must be technically implemented. This includes numerous activities such as detailed design, testing, evaluation, planning of manufacturing processes, and more. All these activities are not addressed in this book, because they are described in biological or product development literature.

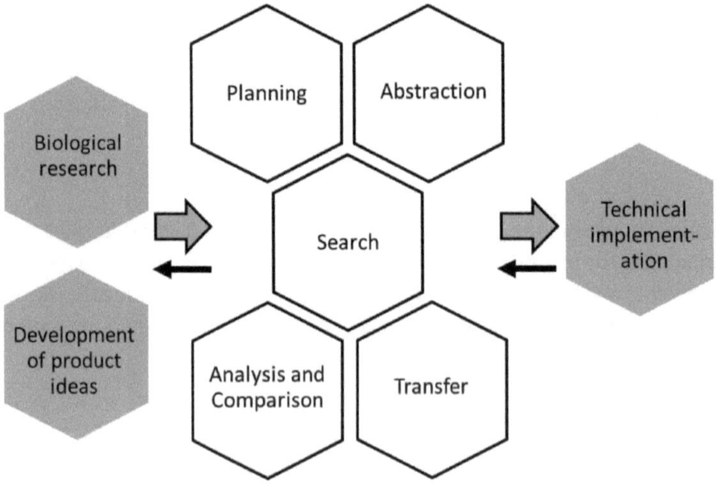

Figure 1-1: Bio-inspired design: specific and other activities

1.2.2 Bio-inspired Design Approaches

In bio-inspired design research, two distinct approaches have been postulated (e.g. by Helms et al. 2009, VDI 6220, 2012): *Technology pull* (also called *top-down* or *problem-driven* approach) and *biology push* (also called *bottom-up* or *solution-driven* approach). These two approaches are described as inverse procedures. Technology pull approaches begin from a technical standpoint – based on a given technical task or problem, a biological inspiration must be found. Biology push approaches originate from a biological standpoint – a known biological solution is to be applied in a technical product.

Even though the majority of bio-inspired design projects and classes focus on the technology pull approach (Shu and Cheong 2014), some researchers claim that the biology push approach is more innovative (VDI 6220, 2012). Helms et al. (2009) observed that designers following a biology push procedure developed multifunctional solutions, but stuck to one biological system. A possible explanation is shown in Figure 1-2:

In the case of a technology pull approach, designers start with a technical task. This entails a predefined set of requirements and desired functions. They stick to these boundary conditions, but are relatively free in considering different biological systems. Therefore, they can be more open to transfer several analogies from different biological systems. A typical example is the bio-inspired bone punch (see best practice example, p. 107): in this project, analogies from two different biological systems (the snake and cat) were transferred to fulfil the predefined requirements of the bone punch task.

If a biology push approach is adopted, designers start with a predefined biological system. They might analyze this system in detail, and they are more likely to restrict their analysis to this system instead of searching for other biological systems. However, they are probably more open to searching for multiple technical applications. Moreover, multifunctional biological systems can inspire multifunctional technical systems. A typical example for a biology push approach is vault-structured material (see best practice example, p. 129). Vault-structured materials have multiple functions, such as stability, diffusion of light, and sound insulation, which have been applied in different, often multifunctional products.

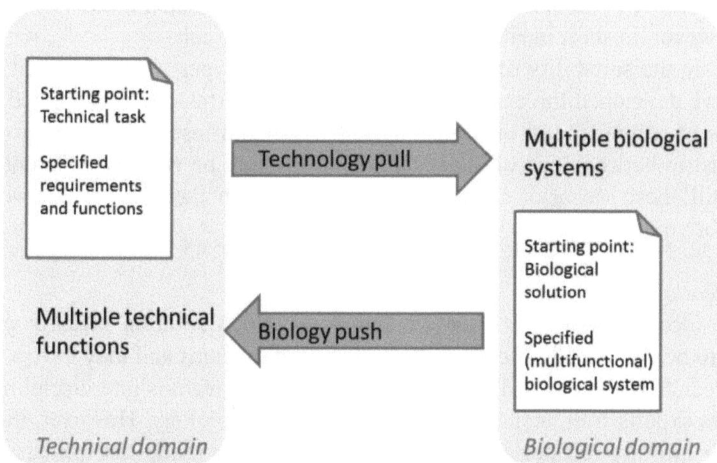

Figure 1-2: Bio-inspired design approaches: technology pull and biology push

To conclude, bio-inspired design projects do not have to strictly follow either a technology pull or a biology push approach. In practice, many opportunities emerge to switch between the approaches and this can be very beneficial. As designers, we should deliberately take a step back and analyze our design process. Considering Figure 1-2, it can be advantageous to question pre-defined requirements and functions (technology pull approach) and to search for several (biological) inspirations (biology push approach).

1.2.3 Bio-inspired Design Scenarios

Having applied bio-inspired design methods, we have noticed that most methods have been developed for specific bio-inspired design scenarios. For other scenarios they may be less suitable. To give an example, the *Causal Relation Template* is a method for formulating a causal relation of a biological system in one sentence and transferring this causal relation to a bio-inspired technical system in another sentence. The method can help designers quickly transfer several analogies from the biological to the technical domain. However, formulating a biological solution in one sentence requires an enormous simplification. If we have conducted a research project to analyze a biological system in detail, we have probably deduced complex explanations for its strategies. To transfer such complex relations, the formulation of a causal relation in one sentence proves less useful.

On the other hand, many methods for bio-inspired design require considerable time and training. For example, *SBF modeling* is a method for abstracting technical and biological systems. The functions of the systems are modeled in detailed steps. This method is very useful for depicting detailed knowledge acquired by literature research or by actual studies or experiments on a biological or technical system. However, in short ideation workshops it is impractical.

To indicate the suitability of the methods for certain types of bio-inspired design, we have developed three scenarios. They differ in terms of effort applied for the analysis of technical and biological system. Your bio-inspired design project might be just in between one of the scenarios, or it might be entirely different altogether. Still, these scenarios can give you a hint as to whether a method is worth exploring for your project.

Creative ideation

Creative ideation addresses short workshops in which solution ideas or concepts have to be developed. They are often conducted in teams and may have a duration between half an hour and several hours. Ideally, the team is interdisciplinary and includes experts from technical disciplines and from biology. However, many methods for bio-inspired design have been developed for mono-disciplinary design teams of engineers. An example for creative ideation is the Mountain bike *BIONICON rEVO* (best practice example, p. 95). A coincidental observation of chamois in the mountains inspired a design team to develop a pneumatic system that enables the user to adapt the bike's front fork angle.

Application of biological knowledge

The application of biological knowledge describes projects with different duration which are typically conducted *from the office desk*. In other words, they usually do not entail biological (or technical) experiments or other studies. They typically start with the rough idea of seeking biological inspiration to solve a technical challenge or with the motivation to apply a biological research result for a technical task. If interesting biological inspirations or technical applications are found, the project often requires a more detailed understanding. Designers then have to search for additional information. In this scenario, the team aspect plays a smaller role, and the designer may be a single engineer or biologist. An example for this scenario is the bio-inspired bone punch (best practice example, p. 107). To develop the bone punch, existing information on two different biological systems (the anaconda and cat) was analyzed. The swallowing strategy of the anaconda and the punching mechanism of the predator teeth of the cat could be abstracted and transferred to develop the bone punch.

Collaborative research projects

Among the described scenarios, collaborative research projects require the most effort in terms of time and project costs. Instead of applying existing biological knowledge, they integrate research into the development project. This is only possible if biologists and engineers collaborate. Typically, such a collaboration results from an unsuccessful attempt to transfer a bio-inspired analogy to a technical solution. To formulate feasible analogies, more knowledge of the biological system is required. However, research in the technical domain can be necessary as well to enable the implementation of a biological analogy with technical means. An example for this is the technical implementation of the Lotus plants superhydrophobic properties (best practice example, p. 69). They result from the Lotus plant's rough microstructure (Neinhuis 2017). To reproduce this roughness on the micro-level of technical surfaces, researchers have developed and tested different materials and micro-elements such as pillars (e.g. Patankar 2004).

1.3 Guiding Example: Frog Adhesion and Application Ideas

Throughout this book, we will use a guiding biological example: tree and torrent frogs and their adhesive toe pads. Additionally, we describe three different technical tasks which could be solved using analogies from tree and torrent frogs' toe pads.

Whenever we want to illustrate a procedure, an approach or a method and explain it on an example, we refer to the tree or torrent frog and one of the technical tasks. We feel that using the same example over and over again can be helpful to understand the essential information quickly. Moreover, the biological system *tree or torrent frogs' toe pads* is very suitable for our purposes, as the toe pads possess multiple properties, implementing several physical principles on different scales to achieve adherence. They can serve as inspiration for multiple technical tasks. As the tree or torrent frog's toe pads allow for the transfer of different analogies, our suggestions for bio-inspired solutions differ; sometimes we propose transferring the hexamer microstructure of the torrent frog's toe pads, other times we propose transferring the abstracted principle of capillary forces, and so on.

We must however emphasize that we use the bio-inspired examples purely for illustrative purposes, and the bio-inspired solutions are only hypothetical. Neither us nor anyone else has developed, tested, or implemented them in reality. They are possibly completely unfeasible.

To show real applications of bio-inspired design, we will additionally present ten best practice examples throughout the book.

1.3.1 Sample Biological System: Tree and Torrent Frogs' Toe Pads

Frogs form the biological order *Anura*, or tailless amphibians. With about 4000 species they account for more than 85% of amphibian species known today. Frogs can be categorized into several families. Frog species from families such as *Hylidae, Rhacophoridae,* and *Agalychnis* are commonly named *tree frogs* because they are often found in trees or smaller plants. Frog species from the family *Ranidae* can be found on rocks in or near running waters, from small streams to waterfalls. They are therefore referred to as *torrent frogs* (O'Shea and Halliday 2001; Faivovich et al. 2005; Klemmstein 2011)

Figure 1-3: Agalychnis callidryas (Costa Rica, Corcovado National Park), photo by Tobias Kohl, Technical University of Munich

Tree and torrent frogs possess remarkable climbing capabilities in wet environments: some tree frogs species can climb up small twigs, some torrent frog species move on wet, slippery, almost vertical rocks. Both tree and torrent frogs possess similar toe pad structures. Adhesive patches are located on the tip of their toes. Additionally, smaller areas can be found on the frogs' feet, in particular tubercles on the toes. Tree frogs use these tubercles for grabbing onto thin twigs. To move forward, tree and torrent frogs "peel off" the adhesive patches from surfaces from back to front. This movement occurs almost effortlessly. Experiments have also shown that tree frogs adapt the position of their feet to the inclination angle of the surface. Sitting on a rotating plate, they position themselves with their head upwards. If the inclination angle increases beyond vertical so that the frogs end up in an overhanging position, they reach outward with their feet to reduce the contact angle between their toes and the surface. When resting on inclined wet surfaces, tree and torrent frogs press their whole body to the surface, probably to facilitate adhesion (Endlein et al. 2013a; Endlein and Barnes 2014; Endlein et al. 2017)

Morphology on different scales

What makes the adhesive toe pads of tree and torrent frogs adhere so well, to both dry and wet surfaces? The answer is multi-faceted; as many biological systems, the toe pads' adhesion is effectuated by several interconnected principles. To understand which principles are predominant, researchers have analyzed the morphology of tree and torrent frogs' toe pads on different scales. Figure 1-4 shows sketches of the scales.

On the integral level, the entire tree or torrent frog and its behavior can be observed. On the macroscopic level, the toe pads are visible.

On the microscopic level, the toe pads consist of layers of cells. The flat surface cells have a width of 10-15 µm. They are mostly hexagonal and they are separated by grooves filled with mucus, which the frogs secrete. The grooves have a width of about 1 µm and a depth of 10 µm. Mucous glands (sketched in black in Figure 1-4) produce the secretion (Drotlef et al. 2015).

There are many differences between tree and torrent frogs. The toe pads of tree frogs possess relatively regular cells, whereas those of the torrent frog have an elongated hexagonal shape. If the frog is in an upright position on a steep surface, the elongation is perpendicular to the direction of gravity forces. Experiments with artificial hexamer structures have shown that in this direction, friction increases (Iturri et al. 2015).

On the submicroscopic level, nano-pillars cover the cell walls, and these pillars are separated by grooves. The tips of the pillars are slightly concave (Endlein and Barnes 2014).

Regarding the properties of material the toe pads consist of, analyses have shown that the outer keratinized layer possesses an elastic modulus similar to that of silicone (5-15 MPa). The softness of this material gradually decreases from the outside to the inside layers. In comparison to other biological tissues, it is relatively soft, but not as soft as adipose tissue.

The mucous film on the toe pads is actually very thin, from approximately 0 to 35 nm. The mucus has a viscosity of 1.4 mPa•s, i.e. 1.4 times the viscosity of water at room temperature. In comparison, honey has a viscosity of approximately 10000 mPa•s, olive oil approximately 100 mPa•s.

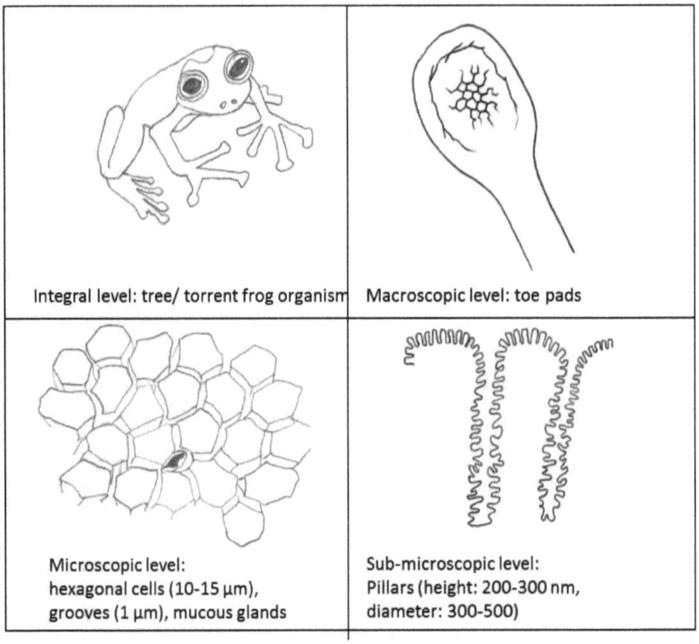

Figure 1-4: Morphology of tree and torrent frogs – hierarchical structure on different scales

Underlying Principles of Adhesion

Today, the underlying principles of adhesion of tree and torrent frogs are still being researched; even though a number of physical effects have been identified, the actual importance and the interrelations of these effects have not been completely understood. In the last decade, research concentrated first on the toe pads' microstructure (e.g. Drotlef et al. 2013; Iturri et al. 2015). Only recently, studies have focused on the nanostructure (e.g. Kappl et al. 2016).

In this book, we summarize all forces under the term adhesion which act against the force of gravity. As Figure 1-5 shows, these forces can be divided into forces acting in parallel or acting at a normal to the surface. Parallel forces prevent the frogs from sliding down the surface, normal forces prevent them from falling off the twig or inclined rock. Measurements of stress values on torrent frogs' toe pads have shown median shear stress values of 3.0 kPa and median normal stress values of 2.4 kPa (Endlein et al. 2013b) for frogs adhering to a dry rotating platform.

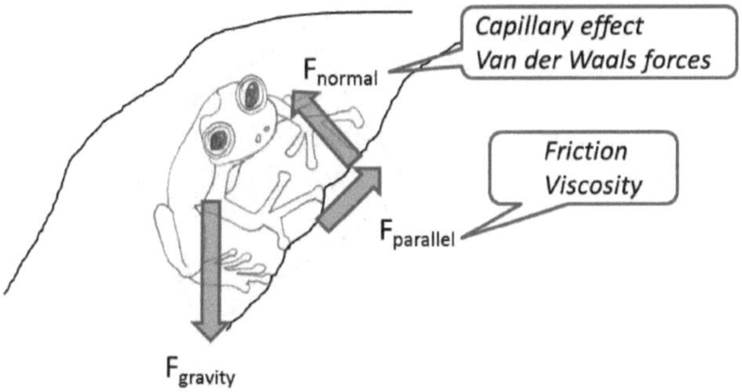

Figure 1-5: Forces and principles of adhesion acting upon a tree/torrent frog

Table 1-1 summarizes the variables which influence the forces acting upon tree/torrent frogs. Friction and viscosity play a role in parallel to the surface:

Friction describes the forces that act against two moving solid surfaces which are in contact. To calculate these "dry" friction forces, a friction coefficient is used, which depends on the surface roughness. Rough surfaces imply a higher friction coefficient than smooth surfaces. Friction coefficients also depend on the state of movement: static friction between non-moving surfaces and kinetic friction between moving surfaces (Hering et al. 2009, p.149–150).

In the case of tree and torrent frogs, the hexagonal cells are covered by nano-pillars. Kappl et al. (2016) have found static friction coefficients of 0.5-1 in immersed condition. In comparison, the static friction coefficient of rubber on asphalt is approximately 0.9 (Hering et al. 2009, p.150). As the nano-pillars are in direct contact with the surface, a probable explanation is that the nano-pillar structure adapts very well to the irregularities of surfaces which increases friction (Endlein and Barnes 2014).

Viscosity implies a different type of friction: the friction between layers of fluids. Assuming a laminar flow, this viscous friction is proportional to the contact area, the velocity, and the viscosity of the fluid divided by the thickness of the fluid layer (Hering et al. 2009, p.213). As explained above, the tree and torrent frogs' secretion has a comparatively low viscosity, similar to water. On the other hand, the thickness of the fluid layer on the toe pad structure is very thin, which increases viscous friction.

Normal to the surface, capillary effect and Van der Waals forces play a role:

The capillary effect describes the rise of fluids in narrow tubes or capillaries. It is induced by intermolecular forces: surface tension of a fluid, and adhesive forces between the fluid and the walls of a tube or capillary. The resultant adhesive force from the capillary effect depends on material properties and on the tube/capillary's geometry. The adhesive force is largest if the tube has a small diameter – trees can transport water from their roots to the tops through capillaries with very small diameters. The adhesive force and the capillary rise are inversely proportional to the tube's diameter (Hering et al. 2009, p.203). As for tree and torrent frogs, both the micro- and nanostructure exhibit grooves. The grooves between the micro-hexamers are relatively narrow (1 µm) and deep (10 µm) (Drotlef et al. 2015). This geometry can induce high capillary forces. Researchers suspect this capillary effect to generate the majority of normal adhesive forces (Drotlef et al. 2013; Endlein and Barnes 2014). The grooves between the nano-pillars possibly have a similar effect.

Table 1-1: Tree and torrent frog's principles of adhesion

Principles of adhesion	Calculation	
Friction	$F_{fr} = \mu \cdot F_N$ µ: static/ kinetic friction coefficient	$\mu_{frogs} = 0.5 \dots 1$
Viscosity	$F_v = \eta \cdot A \cdot \frac{u}{y}$ η: dynamic viscosity $[Pa \cdot s]$ A: contact area u: velocity y: thickness of fluid layer	$\mu_{frogs} = 1.4\, mPa \cdot s$
Capillary effect	$F_c \sim \frac{1}{d}$ d: width of a groove	Tree/ torrent frog's grooves between micro-hexamers: d=1 µm
Van der Waals forces	$F_W \sim \frac{1}{a}$ a: intermolecular distance	Tree/torrent frog's nanostructure allows for distance of a few nanometers

Van der Waals forces describe intermolecular forces (Hering et al. 2009, p.503). They rapidly decrease with the distance between the molecules and are comparatively small. They are only relevant for intermolecular distances of a few nanometers. However, one of the most famous examples for bio-inspiration, the gecko's adhesion strategy, largely depends on Van der Waals forces: the gecko possesses millions of fine hairs (setae). They allow very close contact to surfaces. This results in adhesion forces high enough to allow a gecko to run across a ceiling, upside-down (Autumn et al. 2002; Endlein and Barnes 2014).

As the gecko's setae, the nano-pillars of the toe pads of tree and torrent frogs are suspected to adapt to irregularities of contact surfaces, and to consequently allow for very close contact. This can induce Van-der-Waals forces as well, but to what extent has thus far not been clarified (Endlein and Barnes 2014; Kappl et al. 2016).

Technical reproduction of the toe pad structure

Polydimethylsiloxane (PDMS), a soft elastomer, has been used to mimic the toe pad's microstructure and cover the micro-hexamers' top surface with nanopillars (Endlein and Barnes 2014). Drotlef et al. (2013) have produced PDMS micro-hexamers with flat, t-shaped, and concave tips.

1.3.2 Sample Technical Systems

To illustrate the transfer of analogies from the structure of the tree and torrent frogs' toe pads, we use three different technical examples. The transferred functions and the scale of transfer differ between the examples. As multifunctional systems, tree and torrent frogs' toe pads can serve as inspiration for very different technical applications:

Climbing glove

Climbing gloves are used in outdoor climbing sports. Their main function is to protect the climber's fingers. As well as this, they should provide grip. Therefore, they are often made from cow or goat leather. Cheaper climbing gloves are made from synthetic materials, and these materials are sometimes better for the hand dexterity, which is important for climbing (Beale 2018).

Our hypothetical task is the development of a climbing glove with improved grip. So, let us suppose that a sports goods manufacturer wants to develop a new ground-breaking climbing glove. Due to its improved grip, climbers perform much better. This glove would stand out from the masses of climbing gloves on the market and would probably sell very well. The company's aim is therefore to develop a glove which actually adheres to wet or dry rocks. Of course, adherence does not imply gluing, because the climber must attach and detach their fingers effortlessly.

To conclude, the main requirements for the climbing glove are:

- Improved adherence to wet and dry rocks
- Effortless and quick attachment/detachment
- Protection of the climber's hands
- Dexterity of the hands

Adhesive bandage

Adhesive bandages are very common products. They are applied on small wounds to protect them from environmental influences. To a certain extent, they also absorb blood from the wound. Similar to the climbing glove task, our hypothetical task is to develop a new innovative product with an additional function. In this case, we suppose that adhesive bandages could assume the task of applying a medicine which accelerates the healing of the wound. Then, the user would not have to remove the adhesive bandage to apply such a medicine.

The challenge of this task is to integrate some kind of element into a common adhesive bandage which can emit small amounts of medicine. An obvious location for the application of medicine would be the contact layer of the adhesive bandage. Instead of only absorbing blood, it then also has to store and emit medicine.

The main requirements for the development of the adhesive bandage are therefore:

- Storing small amounts of medicine
- Emitting this medicine, either continuously or several times during the period of use
- Only a slightly larger size than traditional adhesive bandages

Drainage

Drainages collect superfluous water and can prevent inundation. Traditional drainages in gardens consist of precipitous ditches filled with gravel or similar material. Drainpipes in the ditches can drain the water to collection points or to sewerage. However, particularly in dry regions, rainfall occurs infrequently, and, if it does rain, the amount of rain cannot be absorbed by the dry ground. It would be interesting to collect such water in a drainage that does not lead the water elsewhere, but stores and emits the water when the plants need it.

The main requirements of such a drainage task are:

- Absorption of superfluous rain water
- Storage of the water with small losses due to evaporation
- Emission of the rain water

To summarize, all three tasks address the improvement of existing products or technical solutions. They differ in terms of new functions to be fulfilled. The main new function of the climbing glove is adhesion, and for the adhesive bandage and the drainage system, it is the emission of fluids. Moreover, the tasks differ according to the scale of the technical solution. The climbing glove and the adhesive bandage demand a solution on small, most likely microscopic scale. The drainage, on the other hand, requires a solution on a large, macroscopic scale.

These differences in function and scale were our motivation for formulating these three tasks. They are all hypothetical and do not stem from a real industrial demand. The feasibility of the bio-inspired solutions proposed in this book is therefore not tested, and also does not play a role in their illustrative purpose in this book.

2 Planning of a Bio-inspired Design Project

Why might the planning of a bio-inspired design project be an issue? Is it not possible for such a project to be planned as any other product development project would be? Or should it rather be conceptualized as a biological research project?

Bio-inspired design incorporates both technical product development and biological research. There is little crossover between commonly used procedures from both disciplines. Moreover, if engineers and biologists are to collaborate, their different working approaches have to be considered to make the common project successful.

We therefore discuss in this chapter general procedures from both disciplines to highlight potential similarities and differences. We then present procedures which have been developed specifically for bio-inspired design to be used as methods for planning different bio-inspired development projects.

2.1 Technical Product Development Procedures

In technical product development or engineering design, research on product development processes and procedures dates back over fifty years. To understand the differences in the procedures, we must consider their different aims:

A number of prescriptive procedures have been developed as guidelines for designers and design companies. Descriptive procedures, on the other hand, have been developed to understand and explain design (2.1.1). Furthermore, the stages covered by the procedures and the level of detail vary: their focus can lie either on the initial or later phases of product development (2.1.2). Another aspect that differentiates procedures is their sequence of the product development stages and activities – many but not all procedures suggest a linear sequence with possible iterations, whilst others do not (2.1.3).

2.1.1 Prescriptive vs Descriptive

In technical product development numerous procedures have been developed to plan and conduct development projects. Most engineering companies prescribe a specific development procedure that the engineers must follow. One example is the so-called VDI 2221 procedure suggested by the association of German engineers (VDI 2221, 1993).

The procedure consists of several stages: the starting point for the development process is a given technical task. Take, for example, the development of an innovative adhesive bandage. A task could involve developing an adhesive bandage that applies medicine to a wound. The first step requires the designers to explore and define this task in detail (Stage 1). The outcome is a set of requirements which are commonly documented in a requirements or specification list. In the case of this example, requirements have to be defined in terms of the nature and amount of the medicine to be applied. Based on the requirements, so-called functions and sequence thereof are deduced (Stage 2). One function of the adhesive bandage

H. Hashemi Farzaneh and U. Lindemann, *A Practical Guide to Bio-inspired Design*, https://doi.org/10.1007/978-3-662-57684-7_2

would be to protect the wound from the environment's influences; another function to apply the medicine. These functions can be structured and ordered so as to build a functional model, as shown in Figure 2-1. As the figure shows, functions can be divided into sub-functions. For example, the application of medicine entails the storing of medicine and its subsequent release onto the wound. This breakdown of functions into smaller functions serves to simplify the development of a technical solution. It must be noted that, at this stage, the designers must not yet suggest solutions– all functions must be formulated as "solution-neutral". Solutions are instead suggested in the following step (Stage 3): here, solutions are separately devised for the individual sub-functions. In the case of our adhesive bandage example, there exist multiple possible solutions to protect the wound from liquids. Different materials for the adhesive bandage, e.g. impermeable material such as plastics could be used, or perhaps a semipermeable coating that prevents liquids from entering the adhesive bandage from the outside, but allows air to penetrate it. A possible solution for the storage of the medicine could be the use of a sponge material on the inside of the adhesive bandage, which is immersed in medicine during the production process. When the adhesive bandage is applied on the wound, the medicine can be squeezed onto the wound by pressing the adhesive bandage. In the next development step (Stage 4), the product is divided into realizable modules. In the case of the adhesive bandage this could be the outside layer that fulfills the protective function and the inside layer which implements the function of the medicine's application. Then, the modules are detailed separately (Stage 5); for example, the necessary thicknesses of the outside and inside layers are determined. This leads to a preliminary design. By putting the separate modules together and defining their interfaces, the design is completed (Stage 6). The designers in our adhesive bandage example must define the area of adhesive and the area where the sponger material is applied. In the final step, production and operating instructions must be prepared (Stage 7). Here, the user of the adhesive bandage would require instructions on pressing the adhesive bandage to apply the medicine.

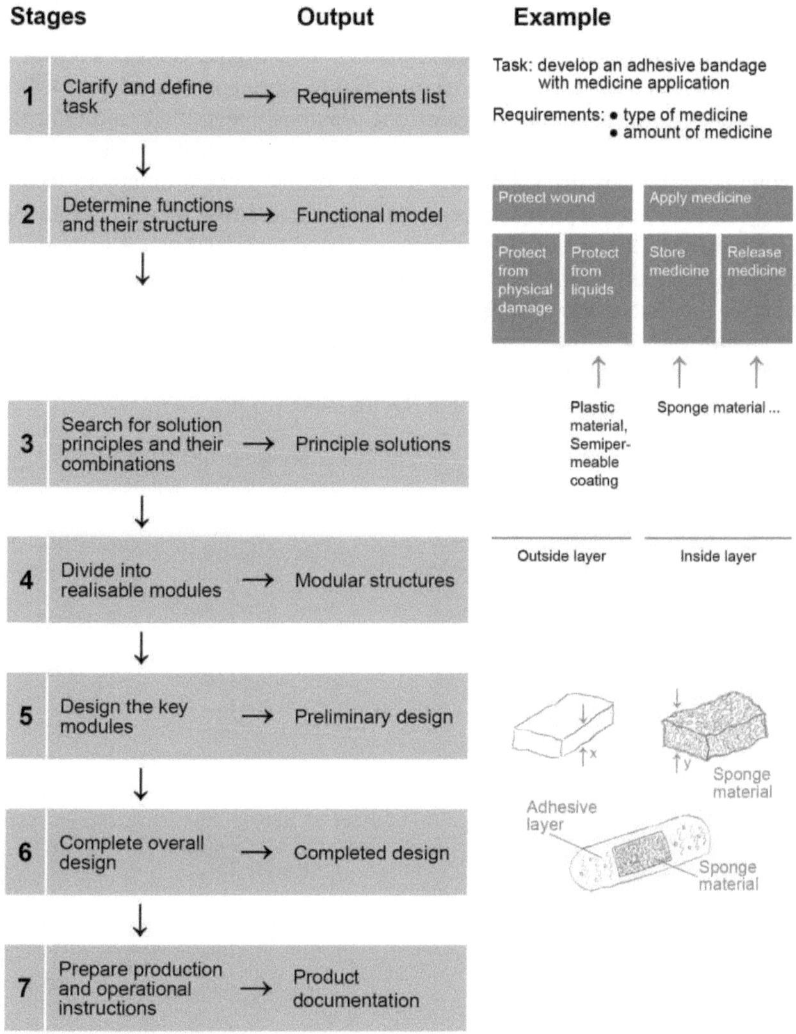

Figure 2-1: Prescriptive technical product development procedure VDI 2221 (VDI 2221, 1993)

Obviously, there is still a difference between the officially prescribed stages of the product development process and the sequence of activities in reality. Researchers have examined and described the reality of designers' and engineers' working procedures. Based on their findings, Cross (2008, p.30) proposes four activities that constitute the design process (Figure 2-2): First, an initial exploration of the design problem – this activity is similar to the first two stages proposed by the VDI 2221 model. Our wound plaster example would see the designer asking questions such as "Which type of medicine has to be applied?" The answers result in requirements and possibly functions, as explained for the VDI 2221 procedure. Based on the exploration of the design problem, the designer generates one or several solutions, such as using a sponge material to store and release the medicine. The solutions are then evaluated against the previously identified requirements. If the solution is not satisfactory, these two activities are repeated. The last activity is the communication of the solution, e.g. as a sketch or as a complete technical drawing for the manufacturing process. To summarize, Cross' descriptive model is less rigorous than the prescriptive model suggested by VDI 2221. The descriptive model does not clarify whether the designers develop an overall solution at once, or whether they subdivide the problem into functions and then either develop them in parallel or one after the other. Research has shown that designers use both approaches to develop a complete and concrete technical solution based on an abstract, incompletely described problem (Dylla 1990, p.139).

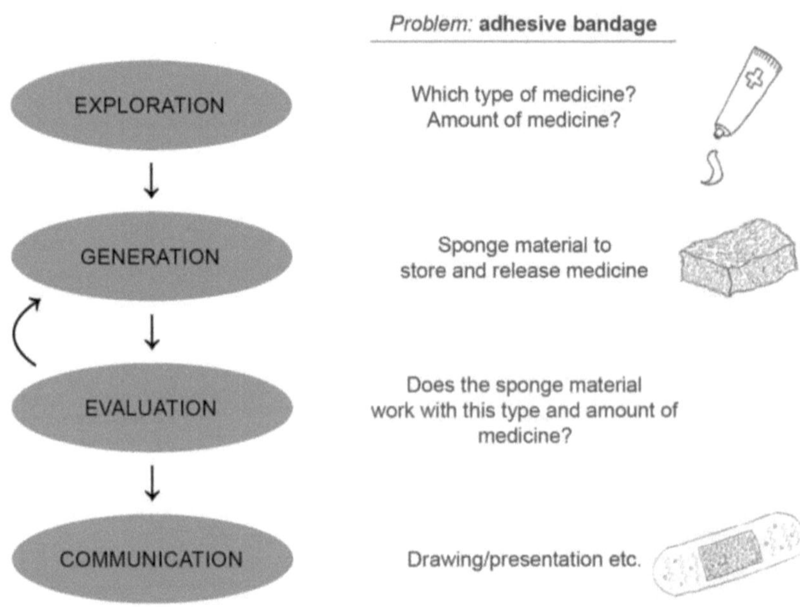

Figure 2-2: Descriptive model of the design process (Cross 2008)

2.1.2 Stages Covered by the Process

An additional aspect of technical product development relevant to bio-inspired design is its scope in relation to company processes. The VDI 2221 model and Cross' model of design cover the core stages of technical product development from an initial problem or task to the presentation of a technical solution. However, they do not explain how the initial problem or task is formulated or how the product is manufactured, used and disposed of.

In particular, the formulation of an initial task or problem can be relevant for bio-inspired design if the starting point is the observation of a biological phenomenon. This stage is specifically addressed by Design Thinking (Brown 2008). Design Thinking proposes three spaces of activities: *inspiration, ideation* and *implementation* (see Figure 2-3). These three stages are not sequential. Rather, designers move between the stages throughout the project. When focusing on the *inspiration* space, activities and methods are proposed. Taking the torrent frog's toe pads as an example, ideas for technical implementation could be explored. One of design thinking's suggestions is to include experts from different disciplines in the product development team. In our example, this could be biologists who are experts on the frog's toe pads, material researchers and engineers. Their task would be to explore possible applications of the structure of frog's toe pads. The toe pads have different functions: They allow the frog to stick to wet, slippery, and vertical rocks. The frog is able to achieve this, in part, by means of the absorption and release of water from the environment. A technical application of the frog's toe pad structure could implement either of these functions: absorption of fluids (e.g. for drainage), release of fluids (e.g. for a medical plaster) or sticking to surfaces (e.g. for climbing gloves). Using the *Design Thinking* approach, the designers could explore the different application ideas in detail. To analyze the business opportunities, they should observe the needs of users, taking also into consideration extreme users, such as children or elderly people. How might they, for example, use or misuse a plaster that releases medicine? Moreover, the project team should explore business constraints related to costs or the reproduction of the frog's toe pad structure with technical material. To communicate and evaluate ideas, *Design Thinking* proposes story-telling and the visualization of the stories in a project room. (Brown 2008).

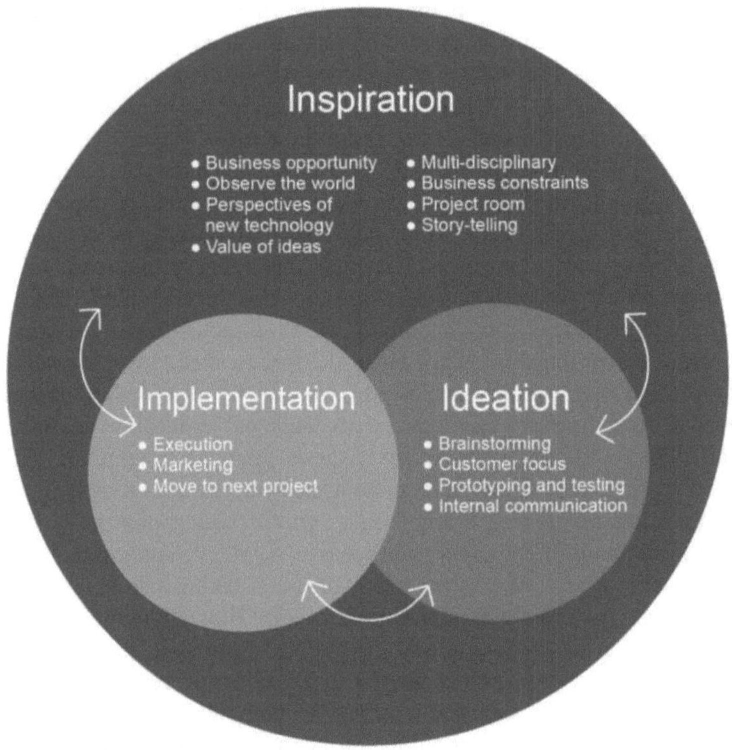

Figure 2-3: Design thinking (Brown 2008)

2.1.3 Sequence of the Product Development Stages and Activities

The models of product development, such as the VDI 2221 or Cross' model, show the designers' procedure as a sequence of stages or activities. This sequence can only be disrupted by iteration if one of the steps is not successful. Other models adopt a less sequential view, showing that product development does not necessarily follow a predefined course of activities and stages. An example is the *Munich Procedural Model* depicted in Figure 2-4. It consists of seven elements which can be traversed in various sequences in addition to the "standard procedure". Moreover, different starting points are possible.

The standard procedure starts by planning the goal of the product development (*Goal Planning*). This goal is analyzed in detail (*Goal Analysis*) before the task is then structured (*Task Structuring*). The structuring of the task can be done by functional modeling as explained for the VDI 2221 model, for example. Next, based on a structured task, solution ideas are generated (*Generation of Solution Ideas*). Unlike the VDI 2221 procedure, the *Munich Procedural Model* emphasizes that the properties of the solutions must be assessed (*Properties Assessment*) before a decision is met to use one or several solutions (*Decision Making*). Finally, the achievement of the predetermined goal must be ensured by assessing possible risks of the solutions and their consequences (*Ensure Goal Achievement*) (Lindemann 2009, p.45–49).

Figure 2-4 shows an alternative procedure through the *Munich Procedural Model*: The starting point in this case is an observation of a biological solution – the absorption and release of fluid by the torrent frog's toe pads. The designers here have the idea of using a copy of the structure of a torrent frog's toe pads for an adhesive bandage that applies medicine to wounds (1). In the next step, they assess the properties of the copied structure's capacity to absorb fluid: the amount of fluid that can be absorbed by the structure is estimated (2). Then, two different materials to build prototypes are chosen (3) and the prototypes' fluid absorption is tested (4). However, in both cases, the amount of fluid absorbed is lower than expected. The designers therefore decide to abandon both solutions (5). To detach themselves from the developed solutions, they reformulate the task of the plaster in abstract functional terms (6). They also formulate the solution of the torrent frog's toe pads in abstract, functional terms before they search for new analogies to solve the task (7). This results in a more abstract analogy to the torrent frog's toe pads: instead of copying the whole micro- and nanostructure of the toe pads, the designers integrate a small medicine-filled pouch into the adhesive bandage. The hexamer surface of the pouch adapts to the body shapes of different users. The channels between the hexamers are porous so that liquid is slowly emitted onto the wound when the user touches the plaster.

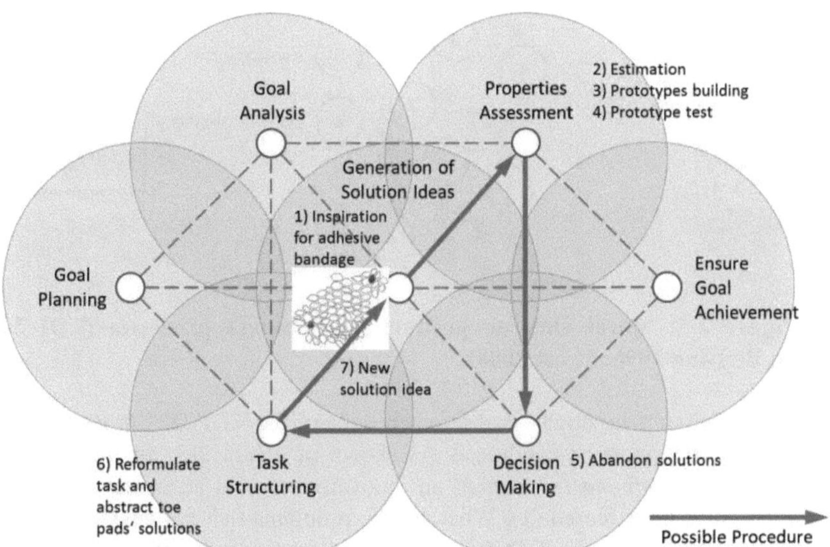

Figure 2-4: Munich Procedural Model (Lindemann 2009)

This example shows that the *Munich Procedural Model* supports alternative starting points (here: *generate solution ideas*), iterations (e.g. from *decision making* to *properties assessment*) and alternative sequences of the procedural steps (from *decision making* to *task structuring*). To which extent the adaption of the sequence is necessary also depends on the type of project for which the Munich procedural model is used: it can be used to plan an entire product development project like the VDI 2221 model, but it can also be used for one procedural step within the project, similar to Cross' model of design. An example for this could be the design of a test bed to evaluate prototypes.

Considering both the macro-level of the entire product development process and the single design steps, Blessing (1996) proposes a spiral view of the product development process, wherein sequential stages constitute the macro-level of a product development project. Each stage is characterized by cyclic, iterative activities which are represented by the spiral. As the solution itself becomes more concrete, the solution space becomes increasingly reduced. This is represented by the concentric shape of the spiral. The spiral view of product development processes is shown in Figure 2-5 integrating the VDI 2221 model and Cross's descriptive model (VDI 2221, 1993; Blessing 1996; Cross 2008).

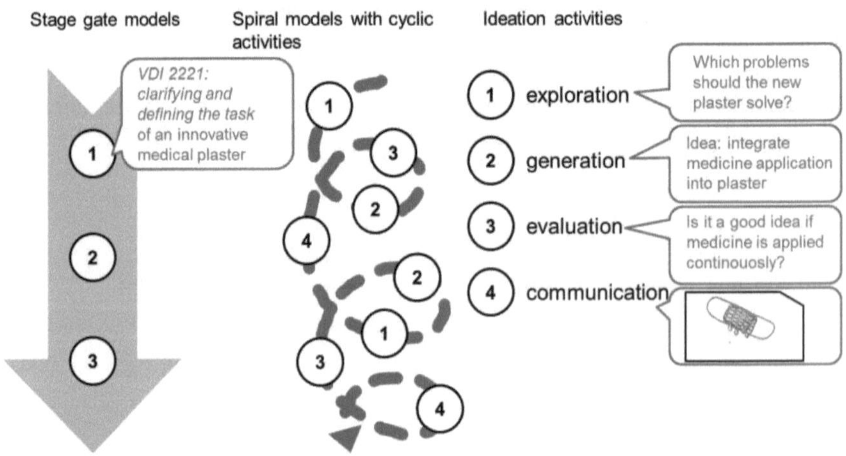

Figure 2-5: Spiral view on product development processes (VDI 2221, 1993; Blessing 1996; Cross 2008)

The spiral view highlights that for each stage of the VDI 2221 procedure, a cycle of activities must be conducted. As shown in Figure 2-5, for the first stage, *clarifying and defining the task* of an innovative wound adhesive bandage, the problem must be explored (1). What are the problems that the new wound adhesive bandage should solve? At this point, ideas for the focus of the product must be generated (2): Of course, the adhesive bandage has to cover the wound. Moreover if the medicine must be applied to the wound, this function could be integrated directly into the plaster. However, it must be clarified whether a frequent or even continuous application of medicine is truly beneficial. At this point, information on the problem must be collected to evaluate (3) and communicate (4) the product ideas, for example by formulating a more concrete task and documenting requirements. Similarly, for the second stage of the VDI 2221 procedure, functions must be explored, generated, evaluated and communicated. These cyclic activities are repeated for each stage of the procedure. However, they are not necessarily explicitly used by the designers. The spiral view on product development processes helps to understand these often implicit cycles of activities (VDI 2221, 1993; Blessing 1996; Cross 2008).

2.2 Biological Research Procedures

As shown in the previous section, the different product development procedures have one thing in common. That is, the starting point is a task or problem and the final result is a product. The aim of technical product development is, of course, to produce a technical product. On the contrary, biological research has a fundamentally different goal. Biology studies life, from the level of molecules to the entire biosphere (Purves et al. 2004, p.2; Campbell et al. 2008, p.1–5). Its goal is therefore to gain knowledge on living organisms. However, gaining knowledge is not a finite process with a defined start and end point. Rather, every piece of knowledge acquired gives rise in turn to new questions. This is a general characteristic of science. Therefore, the biological research process may be described as a cyclic procedure, as the general research procedure proposed by Leedy (1989).

Figure 2-6 shows the details: the starting point is the identification of a problem – the researcher discovers an unanswered problem (1). For example, the reasons for the adhesion of the torrent frogs on wet rocks could be such a problem to explore. In the second step, this broad problem must be narrowed down (2). The problem statement may for example focus on the microstructure of the torrent frog's toe pads. Then, the problem must be divided into sub-problems that can each be answered by one hypothesis (3). Regarding the frog, for example, the microstructure "problem" could be divided into the sub-problems "shape of the microstructure", "deformation of the microstructure due to forces", "absorption of fluid by the microstructure" and "adhesion force due to the microstructure". For each of the sub-problems, one hypothesis is formulated in the following step (4). An example hypothesis would be "The microstructure of the torrent frog's toe pads absorbs fluid on wet surfaces". Facts are then collected with regards to the hypotheses (5), e.g. a microscopic examination of the frog's toe pads could show shape and deformation of its microstructure. These facts are interpreted in order to confirm or reject the hypotheses (6). The confirmation or rejection of hypotheses leads in turn to a new research problem, so that the researcher must start a new research circle. Regarding the frog, discovering the functionality of the toe pads' microstructure would raise questions on the role of their nanostructure (Leedy 1989, p.3–13).

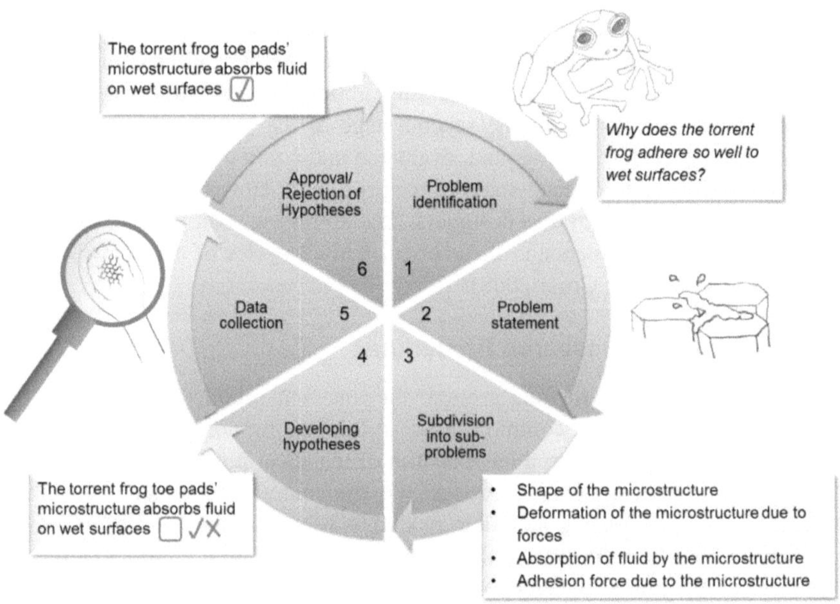

Figure 2-6: Cyclic research process (based on Leedy 1989, p.9)

For biology, Knisely (2005, p.1–9) detailed the general research procedure. As Figure 2-7 shows, she shifts the emphasis from the problem identification, description and subdivision phases to the hypothesis formulation and testing phases. The problem identification phase is termed as "asking questions". Barnard et al. (2007, p.10–12) name four possible sources of questions: curiosity, casual observation, exploratory observation or previous studies. The torrent frog could for example become an object of interest for a biological researcher because of casual observation in its natural habitat. The biologist might wonder why the torrent frog adheres so well to wet and vertical rock surfaces. An additional aspect is the relevance of the question. Questions in biology are relevant if they have a relation to a superordinate theory or general rule (Lamprecht 1999, p.72–75).

In the case of the torrent frog, the examination of its toe pad could reveal similarities to other frog species or other amphibians. This could explain family relationships between the species, for example similar ways of life. In order to "look for sources", the biological researcher conducts a literature study so that he can either answer his question or refine it. Based on the acquired knowledge, the initial question is refined or changed. For example, the microstructure of the torrent frog's toe pads might already be well explored, in which case the researcher can focus on the nanostructure instead. Now he develops possible answers to his questions, also called hypotheses. A researcher could hypothesize that the weight of the torrent frog deforms the nanostructure of its toe pads. The deformed nanostructure possibly increases friction forces and improves the adhesion of the toe pads. To test the hypothesis, data must be collected. The data collection must be planned and the expected outcome has to be predicted.

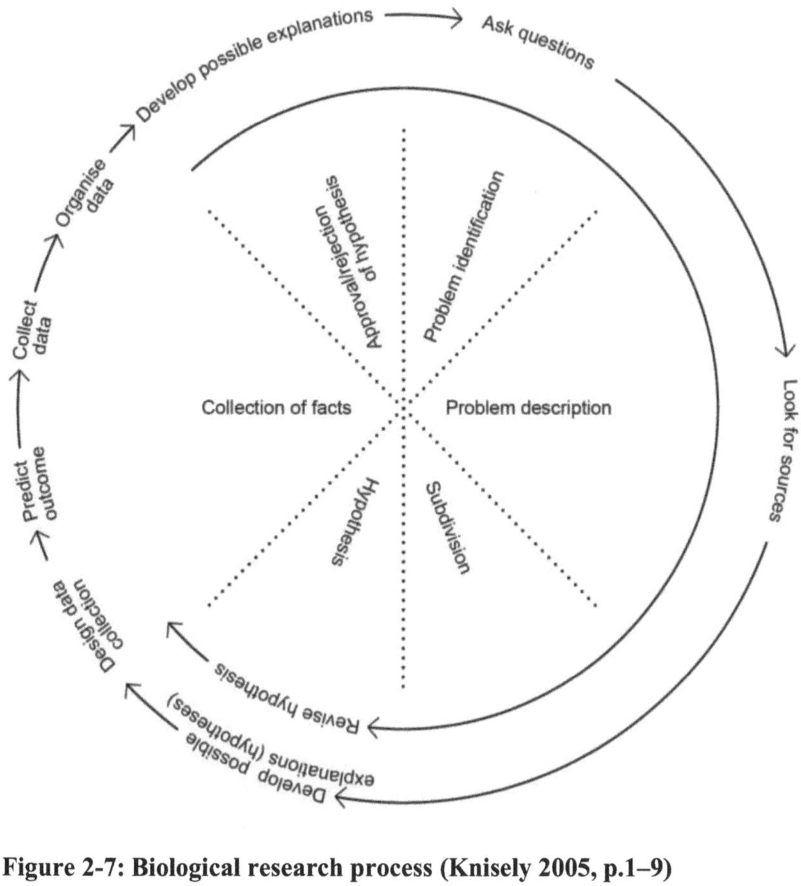

Figure 2-7: Biological research process (Knisely 2005, p.1–9)

As shown in Figure 2-7, the next step of the research procedure is to actually collect the data. The collection of biological data always involves variability. In our example, the area of the toe pads' nanostructure and its deformation due to weight varies between each individual frog. The statistical test supports the researcher in predicting whether the results can be trusted despite the variability. Therefore, the researcher organizes the collected data, applies the statistics and tries to explain the results. If the results do not support the hypothesis as expected, the researcher must reassess his data collection procedure. If human error can be ruled out, the hypotheses may have to be revised. For example, if the conducted experiments do not show an increased area of the nanostructure due to the frog's weight, the nanostructure may play a different role than expected. For example, instead of friction forces, capillary forces may be relevant.

Also, if the original hypotheses are supported by the data as expected, the researcher can formulate new hypotheses based on the acquired knowledge. Using these hypotheses the research can be continued, creating a loop as illustrated in Figure 2-7 (Knisely 2005, p.1–9).

In the following two subsections, we will explain two activities of the biological research process in more detail, namely *developing hypotheses* (2.2.1) and *data collection* (2.2.2). These steps of Leedy's cyclic research process (see Figure 2-6) entail a biology-specific procedure which differs particularly from engineering procedures.

2.2.1 Developing Hypotheses

According to Lamprecht (1999, p.20–21), different types of questions are possible depending on the researcher's perspective. Biological observations can be explained from four different perspectives (Lamprecht 1999, p.20–21):

- Direct causes: The frog might adhere to the wet rock because of a combination of friction and capillary forces that originate in the micro- and nanostructure of its toe pads.
- Ontogenetic causes: The individual torrent frog might have learned to cling to specific places of the rock which have the right angle and the water current of the water fall is not too strong.
- Historical origin: The micro- and nanostructure might have developed from a simpler toe structure of the torrent frog's ancestors.
- Evolutionary causes: The adherence capabilities might enable the torrent frog to live in water falls where predators cannot capture it.

For bio-inspired design, the first perspective – looking for the direct causes of an observation – seems to be most interesting. But it is important to understand that for biologists there can be other relevant perspectives.

A possible hypothesis regarding the torrent frog could therefore be:

"The nanostructure of the torrent frog's toe pads deforms due to the frog's weight so that the contact area with the rock is increased."

At first glance, this hypothesis seems to be very narrow – if we want to understand how a frog adheres to the wet rock surface, why would we formulate a hypothesis on the contact area? When formulating hypotheses, all variables must be measurable. The adherence forces (friction, capillary forces…) might be difficult to measure in experiments with real living frogs. Therefore, a first experiment may only be able to explain the increased contact zone. If the experiments confirm the hypothesis (the contact zone of the toe pads increases), new hypotheses can be set up. For example, a next hypothesis could be: "The increased contact zone of the torrent frog's toe pads increases frictional forces on inclined, wet rock surfaces."

2.2.2 Data Collection

To test the hypotheses, the type of data collection must be planned and its outcome must be predicted. Besides experiments, other forms of data collection are possible. Examples are scientific observation or computational simulations (Kitano 2002; Neugebauer et al. 2004, p.7). Often, a combination of methods is used. For example, Kitano (2002) proposes to use computer simulation for building preliminary models. These models then must be tested in experiments. The procedure for planning data collection differs when comparing computational simulation to experiments or scientific observation. In the case of computational simulation, a model of the biological system must be set up (Kitano 2002; Haefner 2005). The modeling of biological systems plays an important role for bio-inspired design and will be described later in chapter 3.2. For experiments and scientific observation, the hypotheses are instead usually tested using statistical methods. A detailed procedure for planning experiments and scientific observation is shown in Figure 1-8.

In the first step, the scope of generalization must be determined. In the torrent frog example, the scope could be one species of torrent frogs or only a population living at one specific waterfall. Moreover, a representative sample must be chosen. The sample must be representative to allow for generalization. The researcher might, for example, consider torrent frogs of different age, size, weight, and so forth. In the following step, the researcher decides on the type of data collection, i.e. whether an experiment or scientific observation is conducted. At this point, practical issues also must be considered. In our example, a difficulty arising during the experiments might be the availability of torrent frogs for a laboratory observation. One difficulty of observing torrent frogs in their natural habitat might be the ability to measure the nanostructure deformation. The exact measures used for the variables and the measurement methods thereof must be determined in the next step. For the sample hypothesis, the researcher must select a method for measuring the increased contact area between the nanostructure of the torrent frog's toe pads and a surface due to the frog's weight. This could be done using Transmission Electron Microscopy (TEM) (see for example Scholz et al. 2009). Then, the researcher must decide whether paired or independent samples should be used. In our example, a "paired sample" would involve measuring the area of the nanostructure of one frog's toe pads once in an unloaded condition and once loaded with the frog's weight. This sample is paired, because for each frog, two measurements are made, and these measurements are interdependent. The nanostructure's area in the loaded condition must be bigger than in the unloaded condition. Another approach would be to measure the toe pad nanostructure of some frogs in an unloaded condition and the toe pad nanostructure of other frogs in a loaded condition. In our example, this would not make sense, as the nanostructure area of the torrent frogs might differ between the individuals.

After deciding on paired or independent samples, the researcher can choose the statistic test, as it depends on the sample type, the number of samples etc. In the final step, the concrete schedule for experimentation or observation can be planned.

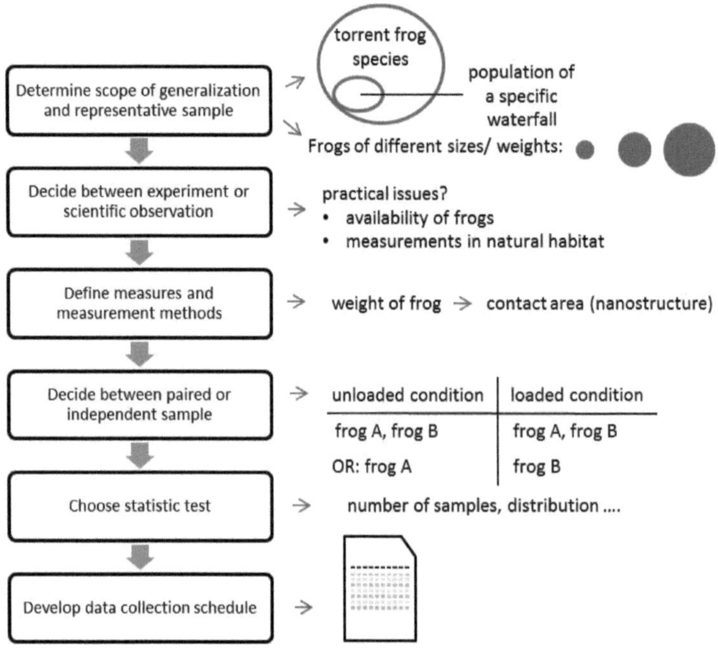

Figure 2-8: Planning experiments and scientific observation (Lamprecht 1999, p.78)

2.3 Comparing Procedures of the Biological and Technical Domain

Table 2-1 summarizes the two preceding sub-chapters, pointing out characteristics of product development and biological research procedures. An engineer or a biologist who wants to develop a bio-inspired product must be aware of the different procedures in both disciplines. Most importantly, both disciplines have different aims: the aim of the technical disciplines is the development of a technical product, whereas the aim of biological research is to understand biological systems. Consequently, the procedure in technical product development is overall linear – there is a defined start and end point. In contrast, biological research is cyclic, new knowledge and understanding of biological systems leads to new biological questions. Different procedural models in technical product development cover different stages of the product development process. Some procedural models focus on the early stages of product development, such as developing a business idea as the goal of product development. Other procedural models detail the late stages of product development, such as the planning of manufacturing. In biology, the data collection methods influence the detailed procedure. For example, planning of experiments considerably differs from planning computational simulation. The stages of technical product development include exploration, generation, evaluation and communication activities. The stages of biological research include different activities, depending on the used data collection method or combination of several methods.

Table 2-1: Comparison of product development and biological research procedures

	Technical Product Development	Biology
Aims	Technical product	Understanding of biological system
Form	Sequences & iterations – overall linear procedure	Cyclic procedure
Different procedural models	Coverage of early or late stages of the procedure	Data collection methods influence the detailed procedure
Stages and activities	Prescription of general stages including similar activities	Prescription of general stages including different activities

2.4 Bio-inspired Design Procedures

Bio-inspired design integrates technical product development and biological research. Consequently, the different approaches to planning and conducting a project in both disciplines must be considered for planning a bio-inspired design project. Still, most procedures that have been developed for bio-inspired design resemble technical product development procedures. In particular, they are not cyclic, but prescribe a rather linear procedure with possible iterations. A possible explanation is that researchers weigh the share of technical product development activities higher than the portion of biological research in a bio-inspired design project. Moreover, researchers in bio-inspired design largely have a technical background. If we follow one of these procedures, especially in long-term research collaborations with engineers and biologists, it can be necessary to integrate biological research cycles into the process. This is shown by example in Figure 2-9.

Here, the *Munich Procedural Model* is used to plan the bio-inspired design project for solving the design task “climbing gloves”. For the activity “generate solution ideas”, the designers and biologists find the biological system torrent frog and its adherence to wet rocks. To understand the frog’s solution principle, they first research biological literature. As there is no complete understanding of the frog’s adhesive capabilities, they plan and conduct experiments with torrent frogs (first research cycle). From these experiments, they hypothesize that the microstructure of the torrent frog’s toe pads plays an important role in adhesion. They use this solution principle for developing possible glove solutions. However, before making a decision on developing one of the alternatives further, they need to “assess properties”. To do this, the designers and biologists build preliminary prototypes with a similar microstructure to the frog and test the adherence forces (second research cycle).

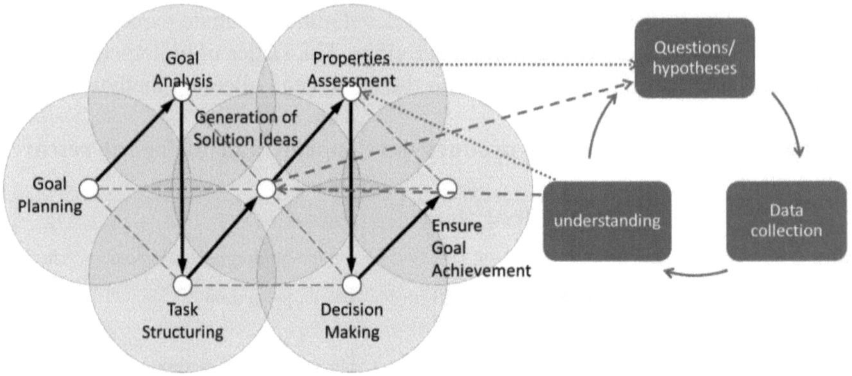

Figure 2-9: Integrating product development and biological research procedures

As in this example, biological research procedures can be integrated into procedures that are based on technical product development procedures. The resulting procedure can be adapted to different scenarios of bio-inspired design.

In this chapter, we present five procedures tailored to different bio-inspired design scenarios and show them on fact sheets. There are a number of other procedures which we will refer to as well, but we have selected these five for the fact sheets as they represent a range of different foci.

To start with, we present one general procedural model that uses a technical task as a starting point: a technology pull procedure developed by Lenau et al. (2010). The next procedural model – a biology push procedure based on Helms et al. (2009) – uses a biological system as a starting point. Then, we present a procedure for collaborative research projects (VDI 6220, 2012). The following procedure proposes a TRIZ-methodology for bio-inspired design (Hill 1997, p.76–77). The last procedure has been developed for the use of bio-inspired design in architecture, civil engineering and industrial design (VDI 6226, 2015).

Figure 2-10 illustrates the suitability of planning methods for different bio-inspired design scenarios. The presented *Biology Push* and *Technology Pull* procedures are suitable in particular for the application of existing biological knowledge – they can also be used for creative ideation, but seem time-consuming in this context. For collaborative research, both procedures provide little support for the research activities. This categorization is even more pronounced for the *TRIZ-based Procedure*. The *Procedure for Collaborative Research Projects* is suitable, in particular, for collaborative research projects both with a biology push and technology pull approach. The *Procedure for Architecture and Industrial Design* fits all bio-inspired design scenarios, yet does not give much detailed support – it can rather be used as a guiding model for bio-inspired architecture, civil engineering and industrial design.

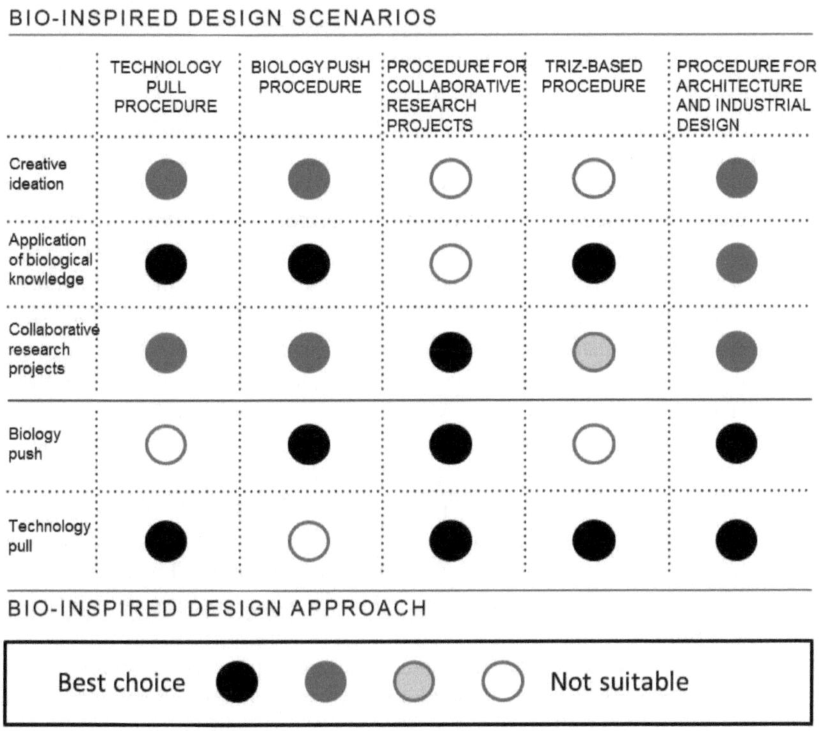

Figure 2-10: Suitability of search methods for different bio-inspired design scenarios and approaches

2.4.1 Technology Pull Procedure

Fact sheet p. 169

Technology pull procedures start with a technical problem or task – they are therefore initiated by engineers or other designers with a technical background. Based on experience from a number of bio-inspired design projects, Lenau et al. (2010) propose a technology pull procedure. They divide the procedure in five main phases. Similar phases have been proposed by other researchers for similar technology pull procedures. Table 2-2 gives an overview:

In the first phase, the “problem definition phase”, the designers analyze the specific technical problem and generalize it. The “generalization” is in fact an abstraction which uncouples the problem from the specific technical context. Taking the development of climbing gloves as an example problem, it can be generalized as a problem of attachment, holding of a certain weight, and detachment. This generalized problem can then be used to generate functional keywords such as attach, hold and detach. As Table 2-2 shows, other researchers propose a similar problem definition phase. For example, the “problem-driven process” includes “problem definition” and “reframing the problem” (Helms et al. 2009).

The next phase is the "search phase". In this characteristic phase for bio-inspired design, the designers use the previously generated functional keywords to search for biological solutions to the general problem. Rummel (2014) propose to "scan nature" in this phase. It must be noted here that not only may functional keywords be used as search terms, but also keywords describing property and environment (more details on this in chapter 4). By performing a search, the designers could, for example, discover the torrent frog, which adheres to wet rocks holding its complete weight on a nearly vertical surface.

In the subsequent "analysis phase", the natural models are discovered (Biomimicry Institute 2016). The biological system and its phenomena are analyzed to extract a biological mechanism. In other words, the aim is to understand how the biological system fulfills the function. In the case of the torrent frog, the designers must understand the main properties of the frog's toe pads that generate the adhesion forces. As we have shown previously, there are multiple "solutions" working in parallel on the micro- and nano-level. In the analysis phase the designers must determine the crucial properties for the adhesion to wet rocks.

The analysis is followed by a "principle phase", which implies an abstraction step (Lenau et al. 2010; Biomimicry Institute 2016): the main biological mechanism must be abstracted into a general functional principle. For example, the torrent frog's nanostructure can be abstracted into a pillar structure with channels that enable a capillary effect.

In the "design phase", the abstracted functional principle inspires a technical idea which can be further developed into an overall concept to solve the technical problem. The technical idea for the climbing gloves can be, for example, to use the nano-channels. This idea can then be developed into a concept using a material with nano-pores for the gloves. The Biomimicry Institute (2016) proposes an explicit evaluation of the bio-inspired technical solution in order to check it against the biological solution (see Table 2-2).

One strength of the *Technology Pull Procedure* proposed by Lenau et al. (2010) is the iteration loops it proposes. If the search phase is not successful, Lenau et al. (2010) propose to go back to the problem definition phase and to redefine the problem. Consequently, different and possibly more adequate keywords for the search can be generated.

If the analysis phase shows that the found biological phenomena are not adequate for solving the problem, then formulating biological keywords can lead to different and possibly more adequate biological systems. The formulation of biological keywords can be helpful because biologists use a different terminology from engineers (Lenau et al. 2010; Helten et al. 2011). A thesaurus can be used to translate technical to biological keywords (for details see chapter 4).

In the principle phase, the designers have to learn more on the biological system in case the abstracted functional principle does not satisfy the desired function, for example if calculations show that the identified capillary effect of the frog's toe pad's nanostructure is not strong enough to hold the frog's weight. Here they would have to study other principles that contribute to the frog's adhesion to wet rocks.

If the abstracted functional principle cannot be transferred to a technical idea, one option is to refine the principles. For example, the scale of the hexamers and channels can be changed. Ultimately, the problem can also be adapted and redefined: concerning the glove example, the aim of the first problem might have been too ambitious, and the weight that the gloves have to hold has to be reduced so that they provide less support (but still at least some support) for the person climbing.

To conclude, the technology pull procedure proposed by Lenau et al. (2010) includes similar phases as other technology pull procedures. It includes practical guidance in the early phases of product development, but few details on the later phases of detailed design of the bio-inspired solution concept. A possible reason is that these phases are not considered specific to bio-inspired design, but rather seen as general product development phases. One limitation is the applicability of the procedure to different bio-inspired design scenarios: the procedure has been developed from a technical perspective for the application of existing biological knowledge by engineers. It gives little support to collaborative research projects with engineers and biologists. To a certain extent, it can also be used to guide a creative ideation procedure – however, in this scenario, it proves to become time-consuming. The technology pull procedure proposed by Lenau et al. (2010) is suitable to be used with the *Biocards* method for abstraction (see fact sheet, p. 178), because the *Biocards* depict the single results of the analysis and principle phase (biological phenomena, biological mechanism and functional principle).

Table 2-2: Overview technology pull procedures

Technology pull procedure (Lenau et al. 2010)	SQUAT (Rummel 2014)	Design Spiral (The Biomimicry 3.8 Institute 2016)	Problem-driven process (Helms et al. 2009)
Problem definition phase	Problem identification	Identify function	Problem identification
	Problem restructuring	Define context	Reframe the problem
Search phase	Scanning nature	Biologize challenge	Biological solution search
Analysis phase	Principle analysis	Discover natural models	Define the biological solution
Principle phase	Principle transfer	Abstract design principles	Principle extraction
Design phase	Modelling of technical solution	Emulate nature's strategies	Principle application
		Evaluate against life's principles	

2.4.2 Biology Push Procedure

Fact sheet p. 171

The starting point of a biology push procedure is knowledge of a biological phenomenon. In contrast to the technology pull approach, the initiator is rather a biologist or a designer with knowledge in biology. As Table 2-3 shows, several similar biology push procedures have been proposed. The biology push procedure chosen for the fact sheet was originally developed to describe observed working processes in bio-inspired design projects (Helms et al. 2009). Nevertheless, it can still be used as a guideline for planning a biology push procedure. Biology push procedures can be assigned to the four general phases shown in the left column of Table 2-3.

Table 2-3: Overview biology push procedures

General phases of biology push procedures	Solution-Driven Procedure (Helms et al. 2009)	Design Spiral (The Biomimicry 3.8 Institute 2016)	Three-Stages Procedure (Nachtigall 2010)
Biological system analysis	Solution identification Define biological solution	Discover natural models	Biological basis: research, describe, evaluate
Abstraction	Principle extraction Reframe solution	Abstract design principles	Abstraction of biological findings: formulation of general principles
Technical system analysis	Problem search Problem definition	Brainstorm potential applications	-
Bio-inspired solution development	Principle application	Emulate nature's strategies Evaluate against life's principles	Technical application: conceptual strategy, comparison of principles, procedure

The first phase deals with the biological system – the starting point is the identification of a biological "solution", for example knowledge of the structure of the torrent frog's toe pads. This broad "solution" is then "defined" in detail, i.e. it is examined in detail so that the strategy used by the biological system is understood (Helms et al. 2009). At this point, the *Three-Stages Procedure* proposes to research, describe and evaluate the biological system (Nachtigall 2010, p.vii–ix). In the case of the frog, this means regarding the toe pads' micro- and nanostructure and the various principles that contribute to the frog's adhesion to wet rocks. One of these principles is a capillary effect which causes the water to be absorbed into the channels between the hexamers on micro-level. On the other hand, these channels emit a fluid which increases viscosity between the fog's toe pads and the rock surface.

As Table 2-3 depicts, the second phase is a phase of abstraction: the biological principles are extracted, i.e. environmental and structural characteristics are abandoned, in order to formulate the principle in a "solution-neutral" manner. The principle can additionally be reframed from a technical perspective (Helms et al. 2009).

In the case of the torrent frog, the designers could list the various forces, such as capillary forces, friction forces etc., as well as their origin. This can help to identify functional chains such as: pressure on the toe pads causing the emission of a fluid through the channels on the micro-level. The shape of these channels enables a capillary effect allowing the fluid to be absorbed.

The abstraction of these biological principles reveals functions – which in the case of the torrent frog would be "emit fluid", "absorb fluid" and so on. These functions can be used in the following phase to search for a technical problem. Helms et al. (2009) divide this phase into "problem search" and "problem definition", i.e. the detailed analysis and understanding of the technical problem. In our example, the designers could come up with the application of medicine as a problem. A further analysis would provide information regarding the characteristics of the medicine (viscosity, stability, etc.), the duration of the application, etc.

This information is necessary for the last phase: the development of a bio-inspired solution. In this phase, the abstracted biological principle is transferred to solve the technical problem (Helms et al. 2009). A detailed bio-inspired solution is then developed, for example a plaster with a hexamer layer releasing medicine onto wounds. Table 2-3 shows that other procedures emphasize the comparison of biological and technical applications (Nachtigall 2010, p.vii–ix), or an evaluation of the biological solution's suitability (Biomimicry Institute 2016) in this phase.

BEST PRACTICE

FlexShapeGripper – A Biology Push Development Project

Festo AG & Co. KG, Oslo and Akershus University College of Applied Sciences, Department of Product Design

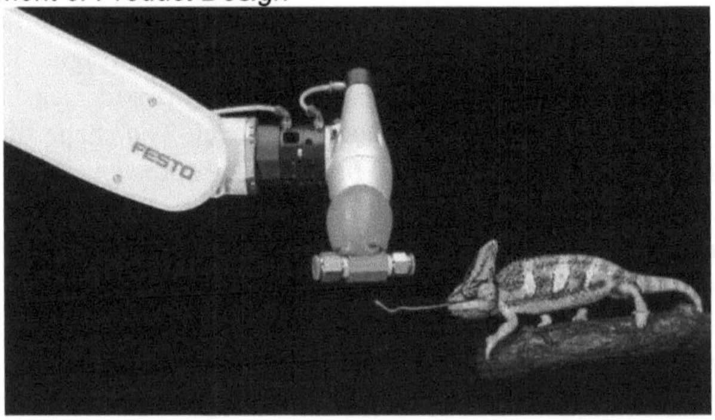

Photo: Festo AG

The chameleon's tongue inspired the *FlexShapeGripper* – an adaptive gripper which can pick, hold and place several differently-shaped objects at once. The development project was initiated by a three-day-workshop with student teams at the Department of Product Design of the Oslo and Akershus University Colleges of Applied Sciences. In the workshop, videos and further information on biological systems was presented to the students. In teams, the students were assigned the task of choosing an interesting biological system and developing a technical application prototype. The chameleon possesses numerous abilities that can be interesting for bio-inspired design; very well-known is its color-changing ability. But, in addition to this, the chameleon also has eyes that give 360° vision, as well as an interesting tongue. To catch prey, the chameleon catapults its tongue at the prey and encloses it. After studying the catapulting ability, one of the student teams took a closer look at this enclosing mechanism. Within four weeks, the team built a prototype using a tube and a jelly-filled cushion to demonstrate the mechanism. Before touching the prey, the center of the chameleon's tongue starts moving backwards, whereas the sides still move forward so as to enclose the prey. In cooperation with the student team, FESTO developed the initial concept into a product within one year. The *FlexShapeGripper* uses a water-filled silicon cap to enclose and pick objects. Pneumatic actuators trigger picking and placing of objects. No forces are needed for holding – the static friction properties of the silicone cap are sufficient.

Based on an interview with project manager at Festo, E. Knubben (September 2017) and Festo AG & Co. KG (2017)

2.4.3 Procedure for Collaborative Research Projects

Fact sheet p. 173

The VDI 6220, 2012 proposes a procedure for collaborative research projects that involve both engineers and biologists. The procedure aims at integrating biologists' and engineers' activities required for the successful development of a bio-inspired product.

The procedure resembles technical product development procedures and focuses on the later phases of the development process. It is based on the development of initial ideas, such that a biological solution must first be matched to a technical task as a prerequisite (either biology push or technology pull). The procedure then details the steps following the initial idea:

In the first step, the biological and technical systems are analyzed. For example, a solution for the application of medicine on wounds is the development goal and the frog toe has been discovered as a possible biological inspiration. In the analysis phase, the frog toe is analyzed using light microscopy. The VDI 6220, 2012 underlines that the methods for analysis have to be adapted to the object of analysis and further proposes chemical, physical and biological methods, as well as methods from material sciences (see chapter 4). The analysis may result in, for example, the frog toe's hexamer microstructure being discovered.

The next step proposes the abstraction of the biological and technical system, i.e. based on the analysis, preliminary relations between the structure, functions, and performance of the systems can be established. This also allows the biological and technical systems to be compared. Regarding the medicine example, it can be hypothesized that the hexamer microstructure absorbs fluid in the channels between the hexamers. In comparison to the technical task, the quantity of absorbed fluid could be similar to the required quantity of medicine. However, this is just an assumption.

In the next step, calculations and/or experiments are used to verify the assumption. Experiments with the biological system may be necessary. But, in our example, an artificial hexamer structure could be built to test the absorption and emission of fluid.

If the assumptions can be confirmed, a technical prototype must be built which also shows whether the bio-inspired product idea can be manufactured. For example, the artificial hexamer structure might have to be integrated into a plaster. This prototype would then be used for application tests. The application tests are used for an overall evaluation of the bio-inspired product. The outcome of the procedure is a bio-inspired invention.

The VDI 6220, 2012 (2012) underlines that this linear sequence of activities is an idealization of the working process – in reality there can be parallel and recursive activities. Nevertheless, a unique feature of the procedure for collaborative research projects is the explicit inclusion of biologists' activities: the analysis of the biological system, the design of experiments with the biological system, and the experiments themselves. On the other hand, the later activities, namely "prototype construction" and "application test" can be considered to rather be engineering-centric activities. Other activities, such as analogy and abstraction are neither predominately one or the other. The concrete mode of collaboration still must be planned by the project leader or team – the procedure does not support the decision as to which activities should be done in collaboratively or by the biologists or engineers alone.

BEST PRACTICE

InspiRat – Interdisciplinary Collaboration in Long-term Research Project

Technische Universität Ilmenau (Biomechatronics), Friedrich-Schiller-Universität Jena (Institute for zoology and evolutionary biology), CAU Kiel (Zoological Institute), Tetra GmbH

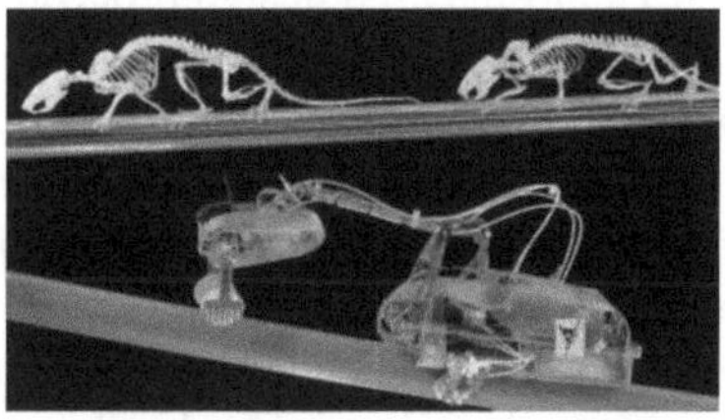

RatNic, Photo: Dr. Andreas Karguth / avateramedical Mechatronics GmbH

RatNic is a bio-inspired climbing robot that can climb in narrow environments and on ropes, wires, and poles. It can be used for inspection purposes in areas difficult to access for humans. However, the project InspiRat had a broader goal than just designing climbing robots. It also provided a systematic understanding of the climbing mechanisms found in small mammals, and demonstrated the applicability of these climbing mechanisms in the technical domain. Moreover, it explored bio-inspired adhesion on different surfaces and its use for robotics. These goals were achieved in a publicly funded, long-term interdisciplinary research project which involved researchers from zoology, biomechatronics, and engineers of an automation company. The project was initiated by the university research groups. At the beginning of the project, discussions and brainstorming sessions supported the understanding of the project partners' views and the generation of initial ideas. Subsequently, the project was divided into four sub-projects according to the goals and the expertise of the various project partners: (1) analysis of the biomechanics of climbing, (2) synthesis of climbing robots, (3) analysis of surface contact, and (4) analysis of bio-inspired climbing. Biologists analyzed the biomechanics of climbing (1), and different species (e.g. rat, squirrel, and chameleon) were observed using x-ray. Engineers developed the climbing robot (2) following a standard technical product development procedure, which included the definition of requirements, component design, overall design, and building and testing of prototypes. Bio-inspired contact and adhesion mechanisms were implemented, tested, and documented in a catalogue (3). Experts in biomechanics integrated the results from the different sub-projects by mathematically modeling climbing, developing a testbed for robots and integrating adhesion mechanisms in robots. The project InspiRat demonstrates that interdisciplinary projects are not linear, but rather must include interwoven sub-projects from biology and engineering. Moreover, they have more than one goal: they can deepen the understanding of biological systems as well as develop bio-inspired technical systems.
Based on an interview with the founder of Tetra GmbH, Andreas Karguth (June 2017) and Fischer et al. (2014)

2.4.4 TRIZ-based Procedure

Fact sheet p. 174

Hill (1997, p.76–77) proposes a bio-inspired design procedure which integrates a number of TRIZ- methods. TRIZ is an innovation methodology which is based on the identification of design principles and other rules in patents (Altshuller 1984). TRIZ methods can help designers to change their viewpoint and find Hill (1997, p.76–77) proposes a bio-inspired design procedure which integrates a number of TRIZ-methods. TRIZ is an innovation methodology based on the identification of design principles and other rules in patents (Altshuller 1984). TRIZ methods can help designers change their viewpoint and find unobvious solutions. Therefore, its adaptation to bio-inspired design has the potential to lead designers to different solutions than other bio-inspired design procedures do.

The TRIZ-based Procedure is a technology pull procedure divided into two main phases: target specification and solution development. In comparison to other procedures, target specification is given particular attention. Hill (1997, p.78) argues that the aim of the procedure is to concretize the technical requirements in such a way that contradictory requirements become apparent. The contradiction of requirements as the core technical problem must be resolved. Hill (1997) supports this by proposing a catalogue of bio-inspirations.

The target specification phase is entirely focused on the technical system: it starts with the study of market situation and customer demands (1.1). This is done by asking the so called "W-questions": *What? Who? Where? When? Why? Whereby? How?* In case of the climbing glove example, the following answers could be given:

What will be developed? → Gloves
Who will use them? → Adult male or female climbers (determines the sizes of the glove)
Where will the gloves be used? → To climb mountains, i.e. there are rocky, muddy, wet or dry surfaces and possibly thorny plants that the gloves must resist
When will the gloves be used? → During the entire climbing / hiking tour, for several hours
Why will they be used? → To protect the hands from injury and wear, as well as additionally facilitate "sticking" to nearly vertical wet and dry rocks
Whereby can the solution be implemented? → The gloves can be manufactured from textiles and coated with synthetic material
How? → The climber will put the gloves over their hands and grip on rocks and other slippery surfaces

Some of the questions may seem trivial and the answers obvious, but this technique helps to reflect the challenges the product has to face. Thereby, it supports by clarifying which technical system to focus on. The next step invovles a system analysis (1.2). With the answers to the W-questions in mind, the designers must draw a map of the technical systems elements and their relations. This structural analysis is shown for the climbing gloves example in Figure 2-11. As the figure shows, the technical system includes not only the product itself, but also the user and the environment. In the subsequent step, a functional analysis clarifies the functional relations between the elements of the technical system. For example, the *thorny plants* can *injure* the climber's *hands*. Therefore, the *climbing gloves* must *protect* them from the thorns.

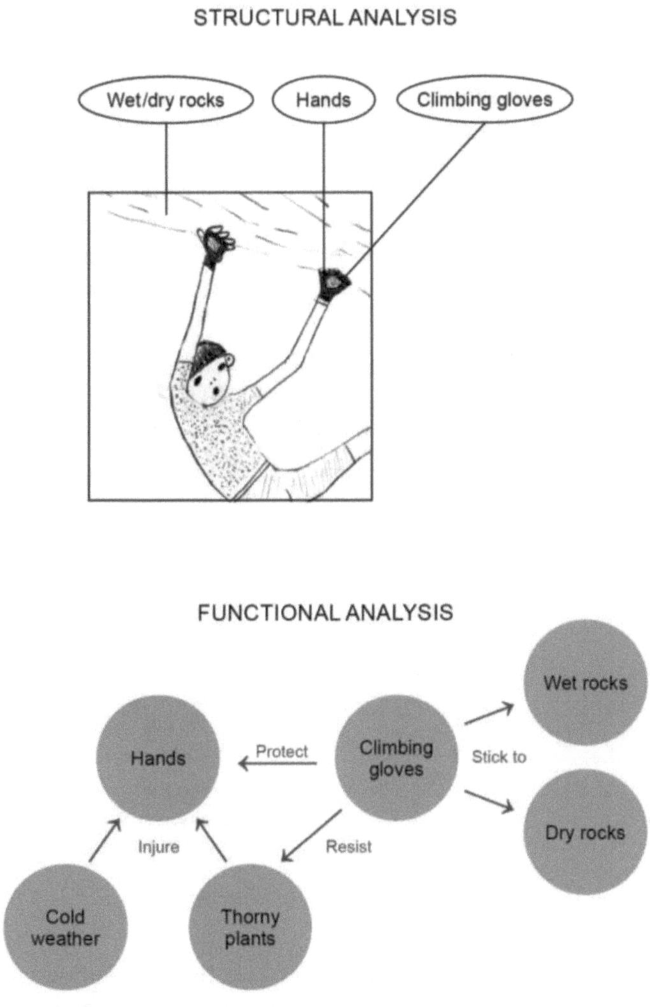

Figure 2-11: TRIZ-based Procedure: structural analysis

Additionally, the *TRIZ-based Procedure* demands an analysis of the state of the art (1.3). Existing products addressing the envisioned product goal or a similar one must be regarded. In the case of the climbing gloves, numerous sorts already exist.

Other products with functions similar to the gloves should also be included. Examples are climbing shoes, suction cups, etc. These products give an idea as to how some of the climbing gloves' purposes can be fulfilled. Moreover, the designers must analyze the weaknesses of these products in order to eliminate them.
In the following step, a generation analysis is conducted (1.4), exemplified by the climbing glove scenario in Table 2-4.
Based on the functional analysis, corresponding properties of gloves are listed. For example, one function of the gloves is to provide protection from cold weather and injuries. The corresponding properties are *temperature/injuries*. Different product generations are identified and analyzed, taking these properties into account. Several thousand years ago, cloth rags served to protect hands from cold temperature. In antiquity, people started to sew gloves from different natural materials. In the 20th century, artificial materials were introduced. *Adherence* to slippery surfaces was improved by gloves made from leather (antiquity) and then in the 20th century by patches from different materials. An additionally identified property is *information*, due to the new development of the "touchscreen compatible" glove fingertips.

Table 2-4: Generation analysis (climbing gloves example)

Functions/ properties	**1st Generation**	**2nd Generation**	**3rd Generation**	**Trends**
Temperature, injuries	Cloth rags (several thousand years ago)	Gloves fabricated by wool, fur, leather (antiquity)	Gloves from a mixture of synthetic and natural materials (20th century)	↑Variety of material ↑Number of pattern pieces ↑Artificial material ↑Multi-functionality
Adherence	-	Gloves from leather (antiquity)	Gloves with patches of "sticky" material (leather/ synthetics etc.) (20th century)	
Information	-	-	Gloves with "touchscreen compatible" fingertips (21th century)	

Regarding the generations of gloves, different trends can be identified, such as the increase in the variety of material used, the number of parts of a glove, and the increase in artificial material. These trends provide a basis for the subsequent step: the determination of the evolution status of the technical system (1.5).
Linde and Hill (1993, p.55–84) propose a number of evolution rules, strategies, stages, and steps that can be used to characterize the evolution status of a product. Especially the evolution stages and rules can support the identification of im-

provement potential of the technical system. According to Linde and Hill (1993, p.55–84), every technical system can be situated in one of the following stages:

Emergence stage: new technology, different technical systems use the technology
Optimization stage: formation of the classical optimized structure
Dynamization stage: adaptation of the classical structure to interact with different environments
Evolution stage: integration of the system into a super-ordinate system

For example, the climbing gloves could be assigned to the dynamization stage: the underlying technologies for gloves exist (material, manufacturing etc.) and there is a typical, optimized structure, i.e. the shape of the gloves. Different gloves have already been developed to adapt to different use scenarios, for example protection against cold, wet, injury, medical applications, and more. However, few gloves adapt to different environmental conditions, such as cold and warm or dry and wet adhesion. In the dynamization stage there still seems to be potential for improvement before gloves can be integrated into a super-ordinate system in the evolution stage.

According to Linde and Hill (1993, p.55–84), the following rules can be regarded to identify potential for further development of technical systems:

Completeness: does the technical system possess a drive, control, transfer, working system and a supporting structure?
Flow of energy: is the flow of energy (forces/electricity etc.) optimized?
Coordination: are the movements/actuation of parts of the system coordinated?
Inexhaustibility: can potential for improvement be identified or can new functionalities be integrated into the technical system?
Ideality: is the effort (e.g. energy/force etc.) required to fulfill the system's functionality minimalized?
Irregularity: are parts of the system underdeveloped compared to others?
Integration: can (parts of) the system be integrated into a superordinate system to improve its efficacy?
Simplification: can the solution principles of the technical system be replaced by simpler principles?
Miniaturization: can macro-principles (e.g. mechanical) be replaced by more effective micro-principles (e.g. chemical)?
Automatization: can the degree of automatization be increased?
Portion of material-field systems: can electric/magnetic/thermic etc. fields be used to simplify a mechanical system?

With regards to the climbing gloves example, the rule of coordination could be used. Current gloves have adhesive patches which do not adapt their adhesive property to the users gripping movement – the adhesive patches currently to not "detach" from the surface when the user withdraws their hand from the surface. Identifying this rule already inspires ideas for actively detaching gloves.

The next step involves gathering effectiveness factors and formulating an effectiveness function (1.6). In this step, the previously conducted methods are reconsidered (area of focus(1.1), system analysis(1.2), state of the art (1.3), generation analysis (1.4) and evolution status (1.5)) to identify effectiveness factors $x_{1...n}$. An effectiveness factor x_n is a target of the development - in the case of the climbing gloves, this could be the adherence to surfaces, protection from injury etc. This effectiveness factor x_n is influenced by technical factors $y_{1...n}$. For example, the adherence to wet surfaces is positively influenced by the contact zone between glove and surfaces. However, in most cases, the technical factors negatively influence other effectiveness factors. A big contact zone negatively influences the formability of the gloves and in turn the climbing speed of the user.

The relevant effectiveness factors are then selected and summarized in an effectiveness function E↑. The target of the development could be formulated in the climbing gloves scenario as:

E↑: (↑grip, ↑climbing speed, ↓manufacturing steps …)

Now, a requirements matrix can be set up (1.7). An initial requirements matrix is shown in Table 2-5 for the climbing gloves example. The effectiveness factors $x_{1...n}$ are listed and the influences of the technical factors $y_{1...n}$ are depicted. The requirements matrix allows for highlighting possible contradictions, so that the designers can select the contradiction estimated to be most relevant.

Table 2-5: Requirements matrix for climbing glove example (based on Linde and Hill (1993, p.89)

Climbing gloves	**Technical factors y**			
Effectivity factors x	y_1: stickiness of glove material	y_2: contact zone	y_3: number of pattern pieces to improve fit of gloves	…
x_1: grip on wet rocks	↑	↑	-	
x_2: climbing speed	↓	↓	↑	
x_3: manufacturing steps	-	-	↓	
…				

Then, based on the selected contradiction, a paradox demand is formulated (1.8). For example, the grip on wet rocks and the climbing speed are inversely related: if the stickiness of the glove material increases, the grip increases, but the climbing speed decreases as the climber has to use more effort to detach the gloves from the rocks.

The formulation of the paradox represents the outcome of the target specification phase: the goal of the solution development is to resolve the paradox and to fulfill the contradictory demands.

In order to resolve the paradox, the first step of the solution development phase is to discover the basic function underlying the paradox demand. Hill (1997, p.77) uses five verbs for the basic functions: *support, transfer, save/block, connect/disconnect,* and *form*. In conjunction with the three conversion products *material, energy,* and *information*, they constitute 15 basic functions. Hill's (1997) use of basic functions is similar to other German design researchers such as Koller (2011) and Pahl and Beitz (2013). The climbing glove example exhibits a paradox that can be related to the function *connect/disconnect material*.

Using the basic functions, the designers can discover biological inspirations listed by Hill (1997) (2.2). The fact sheet *Catalogs/Databases of biological principles* shows more details on this search approach (see fact sheet, p. 191) Examples for biological inspirations listed for *connect/disconnect material* are the squid's suction cup, the lamprey's mouth and the gecko's adhesive toes. More examples are shown in the appendix (section 9.2, p. 248). Collecting these potential biological inspirations in a chart, the designer can now deduce analogies, such as the use of suction cups, similar to the squid, or the use of hair, similar to the gecko (2.3).

The following development steps are then similar to technical product development procedures: the designer has to further develop the analogical solution ideas considering the requirements (2.4). Moreover, he can vary these solution ideas considering size, number, location, shape, material, surface, movement and connection (traction / form fit) of the solution elements (2.4.1). This can be done by using a so-called morphological box. An example is shown in Figure 2-12 for the climbing gloves example. In the morphological box, partial solutions are listed and sorted by characteristics. For example, in Figure 2-12, solutions for adhesion are listed (suction cups, hair) and solutions for the allocation of patches (entire surface/ fingertips/ ball of thumbs). The morphological box supports the designers in finding different possible combinations. For example, suction cups could be used on the fingertips only, or on the ball of thumbs, and so on.

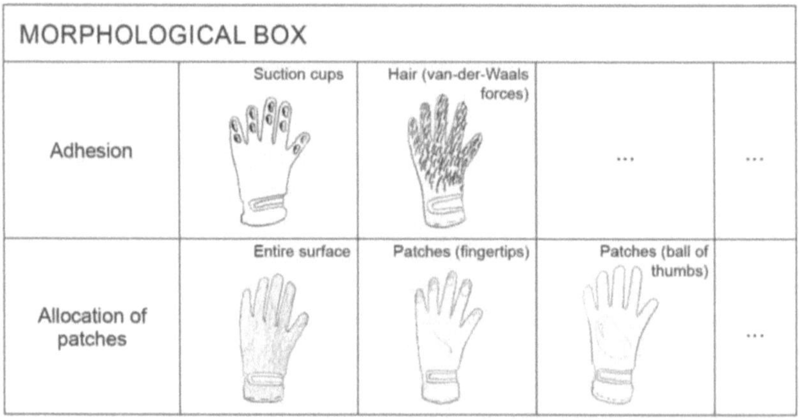

Figure 2-12: TRIZ-based Procedure: morphological box

Solution elements and their variations then must be evaluated to choose the most suitable ones (2.4.2). The evaluation can be conducted by using weighted points, for example. The last step is the detailed design of the technical solution (2.5).

To conclude, the *TRIZ-based Procedure* is a technology pull method that adopts a completely technical perspective. In particular, the formulation of a paradox is entirely different from established product development procedures. Therefore, it can help designers adopt a new perspective and be more creative. The procedure requires effort specifically regarding the target formulation phase and therefore seems to be most suitable for the application of existing biological knowledge, and not for short-term creative ideation.

2.4.5 Procedure for Architecture and Industrial Design

Fact sheet p. 176

The *Procedure for Architecture and Industrial Design* (VDI 6226, p. 12) has been developed to give an orientation for interdisciplinary bio-inspired design in these application fields. One particular aim is to clarify the interplay between semantic and practical function and to exclude a purely aesthetic use of the biological inspiration. The procedure includes two major steps:

In the first step, a "pool of (biological) ideas" is analyzed, for example, by using tables, or morphological boxes. The analysis allows the biological system's semantic functions and practical functions to be identified. Regarding the torrent frog example, an analysis of its toe's microstructure shows the practical function of absorbing and emitting fluid. The knowledge on the practical function must be transferred to a technical invention (patent / utility model). Moreover, the torrent frog has a semantic function related to the practical one: the frog is an amphibian which is closely connected to water. In a symbol-based transfer of impression, the shape of the torrent frog's microstructure, the hexamers, can be transferred to a bio-inspired invention (design registration /copyright / patent / utility model).

In the second step, the resulting product, structure or component ideas are further developed and possibly combined with other solution elements. They can be transformed into a variety of applications, products or systems. In this second step, the use of the practical function is a requirement unlike the use of the semantic function. If the semantic function is abandoned, bio-inspiration is invisible in the final product (bionic-effect-inside). Other options are to visualize the biological system or to use its shape as an integral part. This is the case for the torrent frog example: the toe pads' hexamer structure is enlarged to macro-size to be used for drainage.

To summarize, the advantage of the *Procedure for Architecture and Industrial Design* is the explicitness of the role of semantic and practical function in architecture, civil engineering and industrial design. Within the two broad steps, the procedure remains rather general, giving little orientation as to which concrete methods to use. It can therefore be used in any type of project, from short-term creative ideation to long-term collaborative research projects.

3 Abstraction

Biological systems can rarely be used directly in the technical domain. An example would be the use of purified gastric mucins as a lubrification film for contact lenses (Winkeljann et al. 2017).

However, using biological systems in the technical domain is limited by their complexity if the scale of the system increases beyond molecular level (Sartori et al. 2010). Instead, the essential properties of the biological system must be identified and transferred to a technical solution. The identification and concentration on these specific properties is called abstraction (Prechtl and Burkard 2015, p.6). In both biology and technical product development, abstraction is used to develop models. According to Stachowiak (1973, p.157) models are characterized by three features: firstly, they represent an original (representative feature). Secondly, they constitute a simplification of the original, or in other words, an abstraction (reduction feature). Thirdly, they have a specific purpose at a certain point in time (pragmatic feature). This definition shows that, in addition to the abstraction or reduction feature, two additional features must be considered for modeling:

The representative feature seems obvious at first glance. But, looking closer, one can encounter difficulties regarding the exact nature of the biological or technical system. For example, if we want to model the tree frog scenario: is it only the surface of the torrent frog's toe pads? Or does the behavior of the frog play a role? Or do we have to include elements of the torrent frog's environment – i.e. the surfaces of rocks and plants the frog adheres to? The definition of the system and the system's boundary plays a crucial role for modeling.

The pragmatic feature is linked to the purpose or intended use of the model. The intended use of the model is discipline-specific. Therefore, a model intended to be used for biological research often cannot be used for bio-inspired product development purposes. The pragmatic feature is one reason for the different abstraction and modeling approaches in biology and technical product development.

To explain the different approaches, an overview on the abstraction approaches in technical product development (3.1) and in biology (3.2) is given in this chapter. Based on these approaches, the abstraction methods for bio-inspired design are explained.

H. Hashemi Farzaneh and U. Lindemann, *A Practical Guide to Bio-inspired Design*, https://doi.org/10.1007/978-3-662-57684-7_3

BEST PRACTICE

Prosthetic Hands Inspired by Abstracted Hydraulics of Spider Legs

Karlsruhe Institute of Technology

The movement of spiders has been catching the attention of biologists for almost a century. Spiders possess eight legs, each of which consists of seven segments connected by joints. These joints are flexed and extended by muscles, but some of the joints only possess flexor muscles with no corresponding extensors. How does the spider extend its legs instead? Biologists observed that spiders die with their legs bent. This indicated a possible hydraulic extension of the legs which bend as soon as the internal fluid pressure of the spider collapses. Spiders have an open circulatory system, i.e. their heart pumps fluid directly into the body parts from which it then diffuses back. By this means, spiders can instantaneously fill their legs with fluid, so that they quickly extend. In order to apply this function to technical devices, researchers have modeled spider legs using mechanical theories, such as the theory of curved beams with large displacements. The resulting mathematical models can be used to apply technical parameters (e.g. material properties, hydraulic/pneumatic actuation) and to simulate a variety of technical prototypes. One possible technical application of this type of hydraulic movement is flexible fluidic actuators. They have been used to build prosthetic hands, and, most importantly, they enable highly flexible gripping of objects and very complex movements. Moreover, they are small enough to be integrated into each joint of the fingers of a prosthetic hand.

Based on an interview with designer of prosthetic hands, Stefan Schulz (2017), and Schulz et al. (2001), Vogel, Ferrari (2013; Zentner et al.)

3.1 Abstraction in Technical Product Development

In technical product development, abstraction is recommended as a strategy, for example by Pahl and Beitz (2013). They propose to generalize and simplify the initial technical problem to find a general relation which can be adapted and further developed to a number of technical solutions. Pahl and Beitz (2013) emphasize that this strategy reduces the complexity of the technical problem and makes the important aspects more visible.

How can technical problems and technical systems be abstracted?

Researchers propose a number of abstraction models based on related concepts: An early approach was presented by Hubka and Eder (1988, p.66 ff). They abstract technical systems from the *component* level to the *organ*, *function*, *black box,* and *design specification* levels. On the component level, a technical system consists of physical elements with specific geometries which are linked to each other. For example, a bike consists of two wheels with tires, a bicycle chain, a seat, handlebars and so on. These components are linked by screws, bolts, welded joints, etc. On the component level, technical systems are represented by technical drawings which are used for manufacturing and assembly. An example is shown in Figure 3-1.

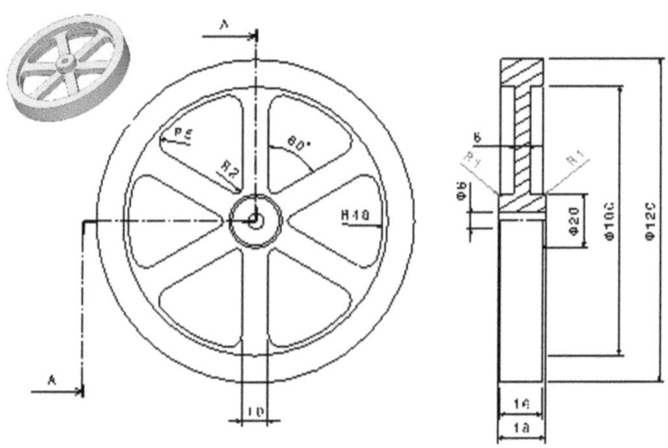

Figure 3-1: CAD model (upper left) and technical drawing of a flywheel

Technical drawings illustrate that the purpose of a technical system is not explained on the component level: even if we can identify some of the components shown in Figure 3-1, for example a screw, we cannot deduce from the drawing neither what the system is used for nor how it is used. Here, the most abstract levels described by Hubka and Eder (1988, p.66 ff) come into play: on the *black box* level, the transformation from an initial state to a desired state is described. If we consider a bike's technical system, the bike's purpose is transporting a person from one location (initial state) to another location (desired state). On the design specification level, requirements are described that must be fulfilled by the technical system. The bike should, for example, enable a certain speed, be tailored to a user with a specific height (child or adult), not exceed a certain weight, and so forth.

How can the transformation stated on the black box level be achieved? This is demonstrated by example in Figure 3-2. On the function level, a composition of functions can be described that enables the transformation of an initial to a desired state: in the initial state, there is the user who can move their legs to walk from one location to another. But, as the speed of walking is rather low, the bike uses several functions to achieve a higher speed: first, it absorbs the user's leg movement (1); then, it transforms the movement to a higher speed (2) and translates this speed back onto the road (3). On the organ level, the implementation of these functions is explained: pedaling translates the user's legs movement to a rotational movement (1), a gear transforms the rotational movement to higher speed (2) and the wheels translate the movement on the road (3). Returning to the component level, the concrete components used to implement the functions are the pedals (1), the bicycle chain (2) and the wheels along with their tires (3). These components are linked via a structure and numerous further components.

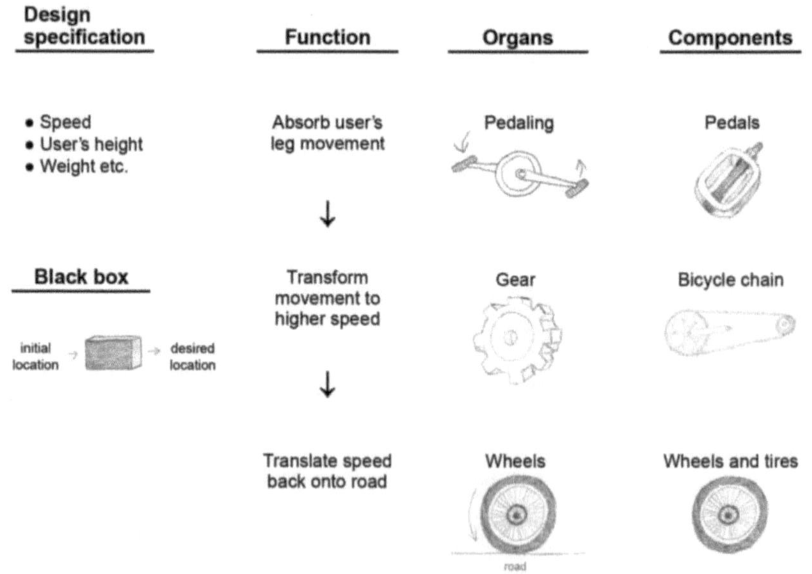

Figure 3-2: Abstraction of the technical system "bike" (based on Hubka and Eder 1988, p.66 ff)

Since Hubka and Eder's (1988, p.66 ff) approach, a number of models for the abstraction of technical systems have been developed. They emphasize different aspects, but the concepts of components, organs, and functions resemble each other.

To summarize, on component level, also called part (Chakrabarti et al. 2005; Andreasen et al. 2014) or structure (Gero 1990; Goel et al. 2009) level, the elements of the physical embodiment of a technical system are described.

On function level, the interplay of actions necessary to reach a desired state or goal is described.

The organ (Hubka and Eder 1988, p.66 ff), behavior (Gero 1990; Goel et al. 2009) or working structure (Ponn and Lindemann 2011, p.27; Pahl and Beitz 2013, p.181) level connects components and functions. On this level, the solution principles that enable the functions are described. They are often physical principles. In the bike example, the third function – translating the speed back on the road – is implemented by using rolling friction of wheels.

How do engineers work with these concepts of abstraction?

Looking back to chapter 2, the abstraction levels determine the procedural stages in most procedural models for planning technical product development. For example, in the VDI 2221, shown in Figure 2-1, the aim of the second stage is to develop a *functional model*. In the third stage, *principle solutions* on the organ level are generated. In the following three stages a detailed design on component level is designed step-by-step: *modular structures*, a *preliminary design* and a *complete design*.

The VDI 2221 and similar procedural models discussed in chapter 2 focus on *new* product development. Their starting points are technical problems or a technical task. However, abstraction can also be used to analyze existing technical problems to deduce improvement potential. This can be highlighted in the *Munich Concretization Model* shown in Figure 3-3:

The *Munich Concretization Model* assembles the abstraction levels of functions, working principles (organs) and embodiments (components) in a so-called *space of solutions*. Parallel to the space of solutions, it highlights a space of requirements which comprises the previously explained design specifications. This representation is chosen to highlight the fact that requirements or design specification develop in parallel to the solutions (Ponn and Lindemann 2011, p.24–28).

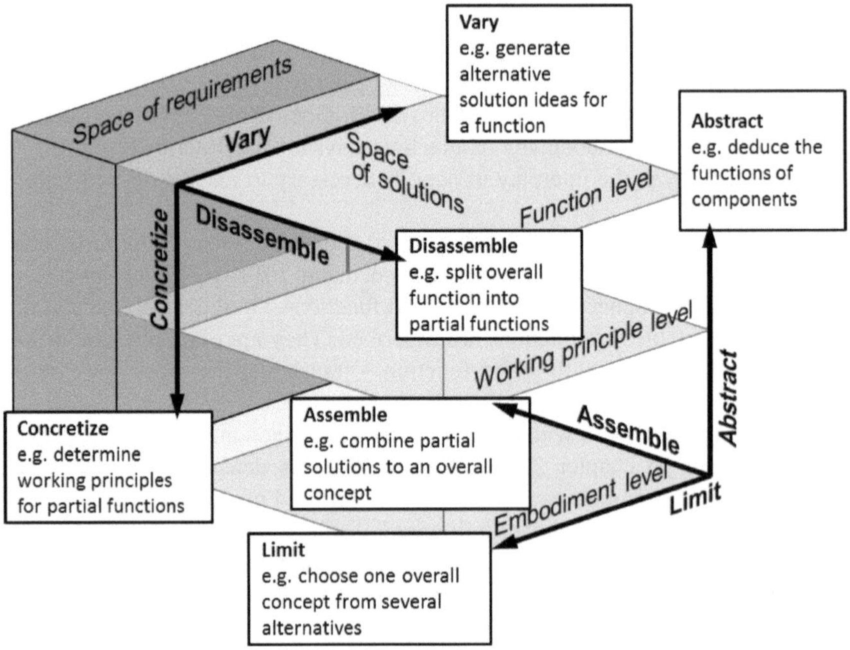

Figure 3-3: Munich Concretization Model (Ponn and Lindemann 2011, p.28)

Even though requirements are collected at the start of a product development project, they evolve throughout the projects as new knowledge of the design problem is generated when designers work on functions, working principles and embodiment. This observation is also called *co-evolution of problem-solution* (Cross 2001).

Within the space of solutions, the *Munich Concretization Model* shows several possible activities:

From function to embodiment level, solutions must be concretized. This is often the case for new product development: the designers search for solutions to a technical problem or task. When bikes were invented, the working principle *pedaling* was chosen to fulfill the function *take up the user's leg movement*. From embodiment to function level, solutions are abstracted. This is the case for the optimization or redesign of existing products. If a bicycle company for example wants to design a cheaper version of an existing bike, all components of the bike can be analyzed regarding their functions. Then, designers can for example think about abandoning redundant functions or designing new components that fulfill several functions.

Apart from the obvious concretization and abstraction activities, the *Munich Concretization Model* displays the adverse activities *vary/limit* and *disassemble/assemble*. In product development, solutions on all levels are varied to generate alternatives. For example, a gear can be implemented by a bicycle chain, but also by cog wheels, belt drives, or otherwise. After creating alternatives, the most suitable alternative must be chosen at a certain point of the development process. This limits the space of solutions. Moreover, solutions can be disassembled, often to simplify the development task and to address the requirements more adequately. For example, the bike's function to transport the user from one location to another can be divided into several functions shown in Figure 3-2. This results in a chain of functions addressing the bike requirements, starting with the function to *take in the user's leg movement*. This chain of function addresses the bike specific requirements. To compare, a car, a motorcycle, and a bus have similar overall functions as a bike: they transport humans from one location to another. By introducing sub-functions, such as absorbing the user's leg movement, the bike-specific requirements (no external energy input!) are addressed more adequately (Ponn and Lindemann 2011, p.24–28).

To summarize, abstraction in technical product development is linked to diverging and converging activities. Numerous methods have been developed, in particular for the elicitation of functions and the development of functional models. They have been used to develop abstraction methods for bio-inspired design. Therefore, we present in the following subsections one method for functional modeling based on a design task, and another method for building a functional model based on an existing product.

3.1.1 Flow-oriented Modeling

Flow-oriented modeling focuses on representing the functions of a technical system or task by material, energy, and signal flows. The functional model is developed based on the technical system's overall function or purpose that is represented by the black box transformation (see Figure 3-2, p. 56). For example, the purpose of the medical plaster example is to apply medicine onto a wound. Figure 3-4 shows an sample flow-oriented functional model according to the approach by Ehrlenspiel and Meerkamm (2013, p.417–426) and Ponn and Lindemann (2011, p.70–74). The modeling approach proposes a circular symbol for the states which are connected via functions represented in rectangular boxes. For the medicine plaster, the material flow can be described by a change of location: At the initial state, the medicine is contained in the plaster material. Via the function *emit material* the desired state is achieved: the medicine is applied onto the wound. To perform the function, energy is needed. In the case of our example the energy is provided as pressure on the plaster. Via the function *touch plaster*, energy is transferred from the user's hand to create pressure on the plaster. A possible signal flow for this example is the signaling of the amount of medicine which has been applied to the wound. For example, via the function *change color*, the amount of medicine applied to the wound can be made visible for the user.

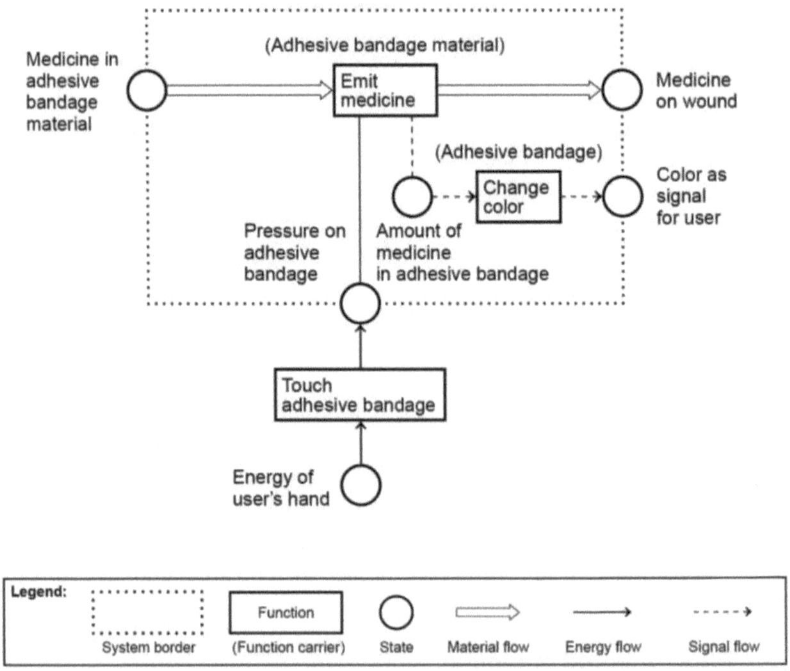

Figure 3-4: Flow-oriented modeling of medicine application task

The flow-oriented modeling approach allows a high degree of abstraction. In its most abstract form, functions can be traced to *generally valid subfunctions* (Pahl and Beitz 2013, p.174). For example, the function *emit medicine* shown in Figure 3-4 can be traced back to the generally valid subfunctions *connecting matter with energy* and *channeling material.* Using generally valid subfunctions, designers can search in design catalogs for working principles fulfilling the functions (for example by Roth 2000 and Koller 2011). However, Pahl and Beitz (2013, p.175) emphasize that in many cases the use of generally valid subfunctions makes the function structure too abstract, i.e. too general to describe the actual challenges of the technical system. For bio-inspired design, it is useful to understand the line of thought of flow-oriented functional models and the use of generally valid subfunctions, as bio-inspired design catalogs have been constructed in a similar manner.

3.1.2 TRIZ Object Modeling

TRIZ is an innovation methodology proposing a multitude of methods for *creative problem solving* (Altshuller 1984). The object modeling is an abstraction method showing the connections between functions and components (Herb et al. 2000, p.101–106). An example for the medical application task is shown in Figure 3-5. In Figure 3-5, the conventional application of medicine and plaster is modeled. The conventional technical system *plaster* consists of three elements: the *carrier material* holds an *adhesive layer* and the *wound contact material*. The *adhesive layer* connects the plaster to the user's *skin*. The *wound contact material* protects the *wound*. It also absorbs blood and other wound liquids. The *medicine* is directly applied by the *user*.

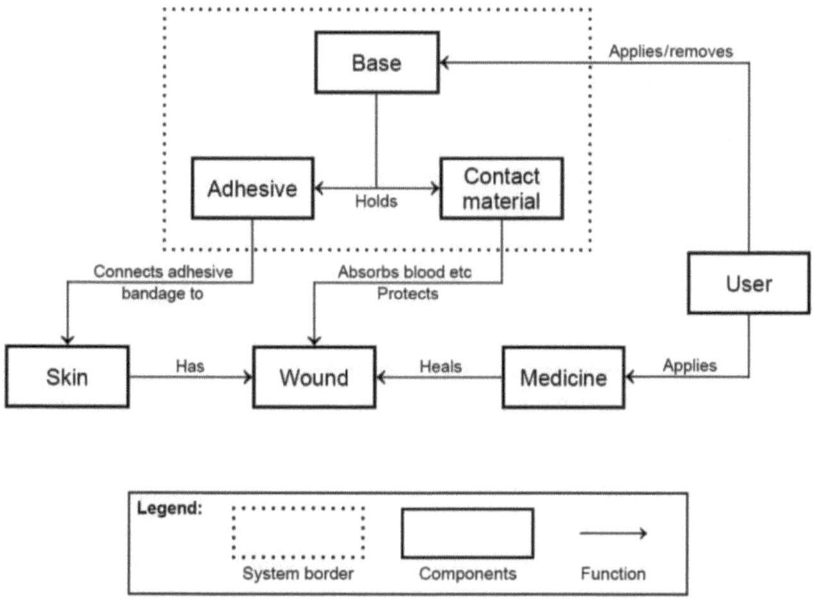

Figure 3-5: TRIZ object modeling of medicine application task (based on Herb et al. 2000, p.101–106)

The TRIZ object modeling gives inspiration regarding how to change the assignment of functions and elements. After setting up the model, the designers should identify components that are expensive or have negative or few connections to the central functions of the technical system. Then, they can follow the so-called *trimming* process, asking themselves the following questions (Herb et al. 2000, p.101–106):

1. Can another component fulfill the component's function?
2. Can the component that is subject to the function perform the function itself?
3. Can the component be eliminated?
4. Can the function be eliminated?

The aim is often to simplify the technical system and reduce costs. In the case of the plaster example, however, the TRIZ object modeling shows that the user must fulfill several functions: they must apply the plaster and the medicine. To apply the medicine, they must remove the plaster and apply another one. An option to simplify the process for the user is to follow step 1 of the trimming process and to assign the application of medicine function to the wound contact material.

To conclude, the TRIZ object modeling is a method for the initial stages of product development to analyze existing products and to identify their weaknesses and potential for improvement and new designs. In this manner, TRIZ object modeling can also be useful for the initial stages of bio-inspired design projects.

3.2 Abstraction in Biology

In the previous sub-section, abstraction was presented as a strategy for the facilitation of technical product development. Is abstraction also a strategy to facilitate biological research? Similar to product development, Leonelli (2008) describes abstraction as an activity for the development of models. Models used in biology can be graphs, and computational models. However, a fact that might surprise biological laypersons is that the term model in biology also refers to so-called *model organisms*. Examples for model organisms are the flowering plant *Arabidopsis thaliana* and the zebra fish. Model organisms are very well researched and can be used to understand general biological phenomena using an integrative research approach (Leonelli 2008; Ankeny and Leonelli 2011; Meunier 2012)

Returning to the general model definition and the three model features proposed by Stachowiak (1973, p.157), models in biology have a representative feature; that is, they can represent biological systems on a level of any size, from the chemical level to the biosphere. Also, the reduction feature is similar to product development models: Leonelli (2008) defines that models are built by selecting specific features of a biological system (a process of *abstraction*). Regarding the different model types such as graphs, computational models and model organisms, different pragmatic features can be identified. The different use of models in biological practice was regarded by Suárez (2007): She analyzed the use of models by one specific research group and observed three types of model use. *Sketches* were used as thinking tools to explain the observed biological phenomenon. *Graphs* represented the relation between two variables measured in several experiments. In the final stage of the research, a *table* was published showing the interpretation and context-independent generalization of work from a number of research groups. Even though Suárez (2007) only considered one case, it is interesting to note that the explanatory sketches were not externally published by the researchers until years after the research problem was solved. This is a relevant observation for bio-inspired design, as it shows that designers cannot expect to find explanatory sketches in publications.

In the following sub-sections, we explain the development of model organisms and computational models in more detail, as they are specific working tools for contemporary biological research.

3.2.1 Model Organisms

"In the most general terms, model organisms are non-human species that are extensively studied in order to understand a range of biological phenomena, with the hope that data and theories generated through use of the model will be applicable to other organisms, particularly those that are in some way more complex than the original model." (Ankeny and Leonelli 2011).

In that sense, several species have been used for biological research, a familiar example being mice and rats used for laboratory experiments. However, the term model organism only emerged in the 1980's. Meinke (1998) argues that in plant biology, before that time research was limited by the small size of research communities. Different research communities worked on similar topics without effectively sharing knowledge. In particular for resource-intensive genome studies such as the Human Genome projects, this changed, and funding for collaboration between researchers increased (Ankeny and Leonelli 2011).

What characterizes a model organism? Table 3-1 shows similar characteristics of the zebrafish and the small flowering plant Arabidopsis thaliana. Both species are considered model organisms (Meinke 1998; Briggs 2002). In both cases, the popularity of the species to be used as model organisms was sparked off by studies related to genetics: Regarding the zebrafish, several studies on mutations were published by different research groups worldwide (Briggs 2002). As for arabidopsis, interest in the plant increased when its small genome was recognized in the 1980s (Meinke 1998). There are a number of practical reasons for using the two species for research: the zebrafish can be easily bred in aquariums (in fact, it is a very popular aquarium fish), it needs short time to reproduce, and the zebrafish embryos are transparent so that all developmental stages can be observed. Similarily, the Arabidopsis plant needs little space and can be cultivated both in greenhouses and under fluorescent light. Like the zebrafish, it has a short doubling time.

For both species, numerous information resources have been funded and established – there are websites, books, and regular conferences. Accordingly, nomenclature conventions have been established. Research results such as information on the genome are available online. Moreover, established conventions on how to breed and experiment with the species have been developed and can also be found online. For the Arabidopsis plant, there is even a seed order service provided by the Ohio State University (The Ohio State University 2017).

Table 3-1: Characteristics of two sample model organisms (zebrafish and arabidopsis)

Model organism	Zebrafish (danio rerio) (Briggs 2002; University of Oregon 2017)	Arabidopsis thaliana (Meinke 1998; Phoenix Bioinformatics Corporation 2017; The Ohio State University 2017)
Practical reasons for use (examples)	Easy to maintain in aquariums, short generation time, all embryonal development stages can be examined in transparent embryos	Small plant, can be grown in petri plates / pots in greenhouse / under fluorescent light, short life cycle (6 weeks)
Initial impulse for becoming a model organism	Mutation studies by several research groups worldwide	Studies revealing the plant's small genome
Information resources	Online information (zfin.org), books, several conferences each year	Online information, books, conferences
Information available (examples)	Nomenclature conventions, genome browser, list of mutants…	Nomenclature, genome browser, publications, database of researchers
Information on cultivation and experimentation (examples)	Breeding, embryonic and larval culture, cellular methods, genetic methods	Order of arabidopsisseeds, detailed instructions on cultivation

To summarize, model organisms are material models used to study a specific biological phenomenon. They are models for whole organisms; research on them is therefore not limited to studying phenomena on smaller levels, such as the cellular level. An important aspect of model organisms is their *experimental scope*, i.e. the generalizability of research results. Results obtained by studying model organisms are more generalizable than results obtained with other experimental ones. However, the degree of generalizability varies considerably depending on the research context. For example, the zebrafish can be used as a model to study the physiology of other fish species, but also to study human genetic diseases (Briggs 2002; Lieschke and Currie 2007).

The amount of research committed to study model organisms provides biologists with in-depth information and knowledge on these species. This is an advantage for further research, but the focus on model organism research has also been criticized because it also limits resources for research on other species (Ankeny and Leonelli 2011).

3.2.2 Computational Models

According to Laubichler and Müller (2007), *modeling and simulation have been the key to most major advances in the biological sciences in recent decades*. They claim that computational models are increasingly used in biological research. Objects of modeling can be categorized according to their dimensionality and complexity (based on Laubichler and Müller 2007):

- Shape and structure of biological system elements
- Interaction between biological system elements
- Development and evolution of biological systems

The aims of computational models are however not restricted to analysis as such, they also include (Laubichler and Müller 2007):

- Big data management (e.g. DNA chips)
- Concept formation (e.g. artificial selection models)
- Theory integration (e.g. in evolutionary development biology, EvoDevo, an integrative biological research field)

The processes of setting up a computational model of a biological system differ from research area to research area. However, Haefner (2005, p.18–24) proposes two general procedures depicted in Figure 3-6. He describes the *classical* modeling process and the alternative modeling process with *multiple hypotheses*.

In both approaches, model objectives are formulated and hypotheses are built. Regarding the torrent frog example, a researcher could have the objective to understand the mechanisms enabling the frog to stick to practically vertical wet surfaces. They then formulate a hypothesis, for example that friction and viscosity are the prevailing forces that enable adhesion. Friction and viscosity can be described in mathematical equations. The actual adhesion of the frog then depends on various factors influencing each other: the inclination angle of the rock, the friction coefficient of the rock, the amount of fluid, and so on. Therefore, the adhesion model cannot be considered analytically. However, a computer simulation can show the effect of the influencing factors. The results of the computer simulation can be compared to real data sets (for example from observation in nature or laboratory tests) in order to evaluate the simulation, the mathematical formulation, and the hypotheses. If, for example, the simulation of friction and viscosity forces results in adhesion forces too small to explain the adhesion of torrent frogs on nearly vertical wet rocks, the researcher first must check the simulation parameters, the underlying algorithm, and the mathematical formulation. If these are correct, the hypotheses are probably wrong, i.e. there are possibly other forces other than viscosity and friction that enable the frog's adhesion to wet rocks. For example, the capillary effect might play a role for the torrent frog's adhesive capacities.

If the researcher follows the classical approach, alternative hypotheses are built sequentially, i.e. only when one hypothesis has been discarded. If the researcher works with multiple hypotheses, these are tested in parallel so that the best hypothesis can be chosen comparing the simulation results.

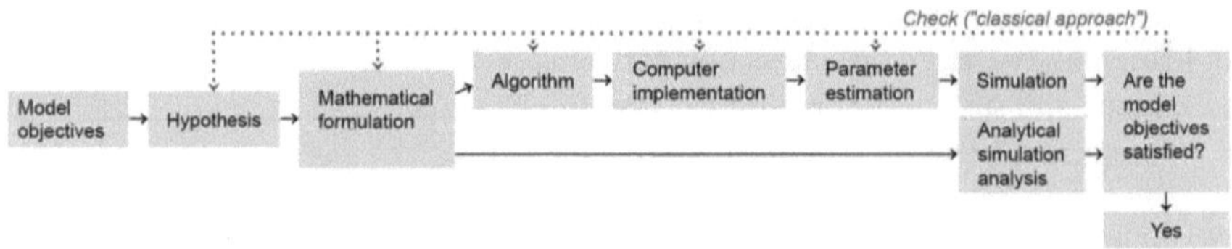

Figure 3-6: Setting up a computational model (simplified depiction of Haefner 2005's approach)

Regarding Haefner's (2005, p.18–24) procedure for setting up a computational model, the biggest challenge of the process is the mathematical formulation of the hypotheses. He proposes the use of so-called *Forrester diagrams* to tackle this challenge. Originally, Forrester (1973, p.130) developed the diagrams to model industrial system dynamics. *Forrester diagrams* are diagrammatic representations of systems showing system objects or states and the relations between them.

This is shown in Figure 3-7 for the torrent frog adhesion example. The waterfall and the frog toes are modelled as sources or sinks represented by clouds. This means that they are considered outside the system and are not explicitly modelled. In the case of the torrent frog, the toes are of interest to the researcher, but modelling them entirely is too complicated for setting up a computational model. Therefore, only the fluid emission and absorption of the toes are modelled by *rate equations* (triangular representation). Both emission and absorption determine the fluid amount which is considered a *state variable* (box representation). The fluid amount serves as an input for calculating the fluid layer (rate equation). The fluid layer influences the friction coefficient. Another influencing factor for the friction coefficient is the water which is dispersed on the rocks. Using the friction coefficient, the friction (rate equation) between frog and rock can be calculated. It determines the adhesion force. The actual influence of state variables is represented as information flows (dashed arrows). There are additional variables which influence the rate equations. Unlike the dynamic state variables, these variables are assumed to be constant. In our example, the frog weight is such a variable. The frog weight is assumed to have a specific constant value, for example the average weight of a frog population – we do not want to model the influence of different frog sizes and weights. After setting up the Forrester diagram, the rate equations can be formulated and the variables can be quantified – the mathematical formulation is completed.

The Forrester diagram described here is simplified, Haefner (2005, p.32–53) explains several other elements which can be used in Forrester diagrams. However, Figure 3-7 gives a general idea of the elements and use of Forrester diagrams.

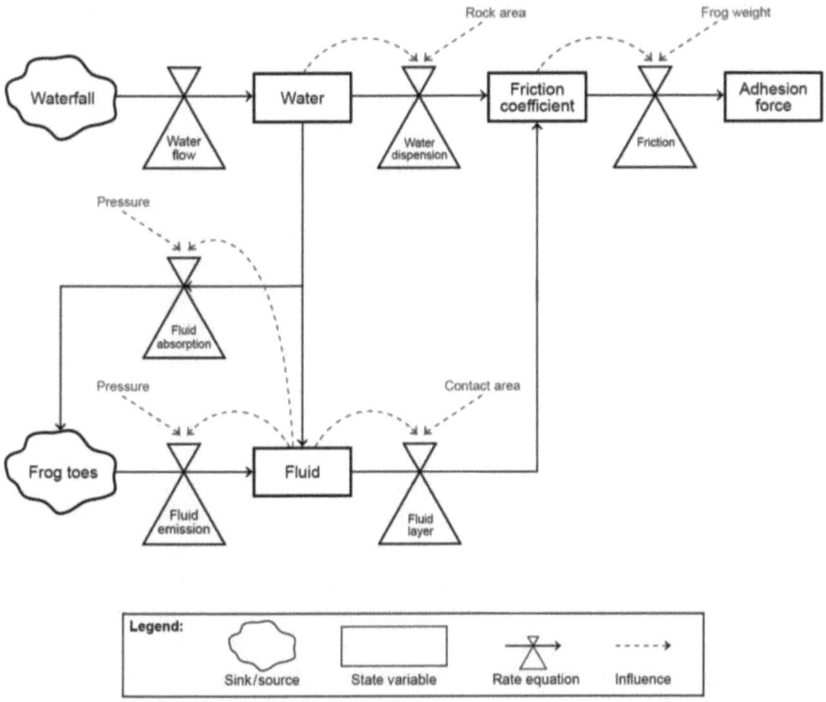

Figure 3-7: Forrester diagram viscosity and friction influence on torrent frog adhesion (following Haefner 2005's approach)

To conclude, computational models have been increasingly used in biology over the last decades. Computational models are based on mathematical formulations and therefore generalized and accessible without specific biological knowledge. The can thereby provide a bridge for a better mutual understanding between the disciplines.

3.3 Comparing Abstraction in the Biological and Technical Domain

In the previous sub-sections, we have explained different types of abstraction used in product development and biological research and the resulting models. The sub-sections, however, do not provide comprehensive accounts of every abstraction. They only highlight typical and discipline-specific approaches. For comparison, Table 3-2 summarizes similar abstractions used in both disciplines.

The table is tentatively divided into abstractions which are estimated to be used predominantly internally (company-internal, laboratory-internal) and abstractions used both in internal and external (e.g. publications, patents) context.

The abstractions that are used predominantly internally are sketches and network models. They are not commonly used to showcase results of product development and biological research externally, but seem to be used mostly as working or thinking tools (Suárez 2007).

As for sketches, designers sketch solution concepts and detailed designs, for example for problem solving, creativity, and communication (Pache 2005, p.116). Similarly, biologists use sketches as thinking tools (Suárez 2007).

The product-development-specific functional models can be considered network models, consisting of elements and their relations. Forrester diagrams are also network models, but their context of use is very different: functional models are used for structuring a technical problem or system (Ponn and Lindemann 2011, p.70–74 Herb et al. 2000, p.101–106), whereas Forrester diagrams are used to prepare a mathematical formulation of a hypothesis (Haefner 2005, p.32–53). Nevertheless, flow oriented functional modeling in particular (see sub-section 3.1.1) follows a similar pattern of logic as Forrester diagrams.

The model organisms (Ankeny and Leonelli 2011) used in biological research (see sub-section 3.2.1) are a type of physical model. Physical models are also used in technical product development, for example as physical prototypes used to test functionality or to show visual or haptic properties of a product. Still, the nature and use of physical prototypes and model organisms are fundamentally different.

The use of graphs to represent measurement data is very similar in both disciplines. The use of mathematic equations also has similarities. In both disciplines they can be used to build computational models. However, biological systems are much more complex than technical systems and the physical laws governing a process are therefore often hypothetical.

Table 3-2: Comparison of selected product development and biological research abstractions

Abstractions	Product development (examples)	Biological research (examples)
Predominantly internal use		
Sketches	Sketches of solution concepts and more detailed designs	Sketches as explanations of biological phenomena
Sequential/ network models (elements, relations)	Functional models (e.g. flow-oriented, TRIZ)	Forrester diagrams
internal and external use		
Physical models	Physical prototypes, test beds etc.	Model organism
Graphs	Data acquired from measurements (e.g. experiments, simulation)	
Mathematic equations	Physical laws, assumptions (≈ hypotheses) regarding solution concepts	Hypotheses (e.g. for computational models), physical laws

BEST PRACTICE

The Lotus Effect – How the Right Model Paved the Way for Success

University of Bonn

The Lotus Effect – today, it is probably one of the most known bio-inspirations. However, its success story started only about twenty years ago. For years, biologists had analyzed the surface structure of the lotus plant's leaves. Initially, their aim was not to understand its superhydrophobic properties, but to examine family relationships with similar plants. Nevertheless, the research provided huge amounts of data on the microstructure of the leaves. The biologists began to wonder whether this structure and its superhydrophobic properties might not be interesting for technical applications. They presented the lotus to several engineering companies. All of them showed interest but finally were discouraged as they were unsure how to mimic the biological structure by technical means. No company wanted to invest money in studying the lotus effect and its applicability in technical products in more detail. The biologists then decided to provide an abstraction and build a simplified technical prototype. They had not only observed that the lotus plant leaves were hydrophobic, but, on the micro-level, the surface had cone-shaped elevations. These cones cause water to form spheres. The spheres roll from the leaves and take dirt particles with them. The biologists mimicked this micro-roughness of the lotus leaves by gluing PTFE particles onto one side of a hydrophobic plate, and then put red powder on it. To demonstrate the effectiveness of the lotus effect, they rinsed the plate with water. The side with the PTFE particles was instantly clean whereas some of the powder stayed on the smooth side. Even though this prototype was not stable, it convinced engineering companies to further research the lotus effect. Today, the lotus effect is used in numerous products such as wall paint, and cleaning and sealing products.

Based on Barthlott et al. (2017), Neinhuis (2017)

3.4 Abstraction Methods for Bio-inspired Design

Most abstraction methods for bio-inspired design have been developed from a product development perspective: they were developed based on the concept of functions as a central element. Nevertheless, all abstraction methods are suitable to model biological systems and use a step-by-step method to facilitate the abstraction. As far as the different abstraction types shown in Table 3-2 are concerned, the bio-inspired design abstraction methods use the predominantly internally used abstraction methods such as sketches and sequential or network models.

Figure 3-8 visualizes the suitability of the presented abstraction methods for different bio-inspired design scenarios.

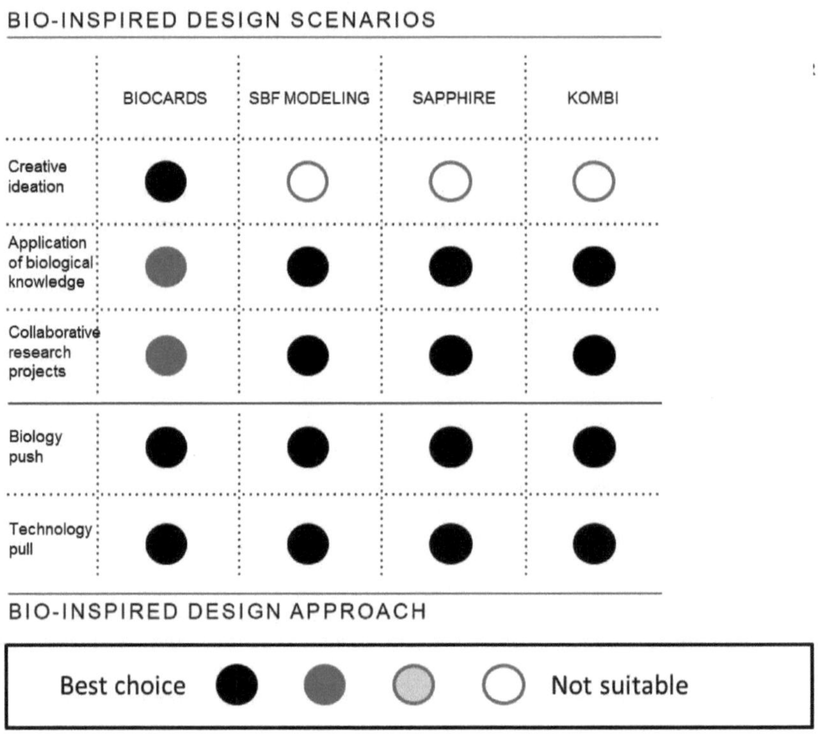

Figure 3-8: Suitability of abstraction methods for different bio-inspired design scenarios and approaches

Biocards (Lenau et al. 2011; Lenau et al. 2015) is a step-by-step abstraction approach for biological systems. *Biocards* use textual descriptions and sketches for describing the biological system. They are most suitable for the preparation of creative ideation. In creative ideation workshops, *Biocards* can provide simplified and, via the visualization, easily accessible information on biological systems. They can also be used in other bio/inspired design scenarios, but their degree of simplification poses a limit. SBF modeling (Goel et al. 2009; Yen et al. 2014) and

SAPPhIRE (Chakrabarti et al. 2005) are more suitable for this context, as they explain biological and technical systems in detail. The necessary expertise and the effort to use these abstraction methods make them less useful for creative ideation. In terms of level of detail, *KoMBi* (Hashemi Farzaneh et al. 2015) is in between the previous approaches. In contrast to them, *KoMBi* also incorporates aspects from biological models.

3.4.1 Biocards

Fact sheet p. 178

Biocards (previously called inspiration cards) are an approach for abstracting biological systems and their phenomena (Lenau et al. 2011; Lenau et al. 2015). They have been developed from a product development perspective. Nevertheless, Lenau et al. (2011) describe a case where *Biocards* have been formulated in interdisciplinary collaboration. The principle idea of *Biocards* is a step-by-step abstraction from the concrete biological system and phenomenon to an abstracted functional principle. As we have observed previously in this chapter, functions are a technical concept – therefore, *Biocards* provide not only an abstraction but also a transformation of the biological system to a technical viewpoint. Visually, *Biocards* include textual descriptions and a sequential model illustrating the functional principle.

The fact sheet depicts a sample *Biocard* describing the adhesion of tree and torrent frogs to wet and dry surfaces. Following the step-by-step approach, the phenomenon (*adhesion to wet and dry surfaces*) and the biological system (*tree or torrent frog*) are firstly named. A sketch or photo shows the biological system. The phenomenon and its underlying principle are briefly described in one sentence using biological terminology.

Then, the biological mechanism is described in more detail using technical terminology. In the torrent frog example, we can describe the structure, deformation and fluid emission/absorption of the toe pads on the micro- and nano-level. In the following step, the underlying physical effects are identified to formulate the abstracted functional principles. As for the torrent frog, there are several physical principles, for example the capillary effect of the channels between hexamers (micro-level) and nano-pillars (nano-level). The depiction of the functional principle illustrates the textual description.

Biocards have several advantages. To start with, they provide an easily accessible and brief overview of a biological phenomenon (they are restricted to one page!). Their brevity also reduces the preparation effort. Moreover, the step-by-step approach facilitates abstraction.

As they do not prescribe a rigid terminology or formalization, *Biocards* leave a lot of freedom as to how they may be used. For example, Lenau et al. (2015) examined the effect of using *Biocards* with functional principle descriptions closer to the biological system to more abstract descriptions. The example on the fact sheet is highly abstract; Figure 3-9 shows a more biology-close description. Lenau et al. (2015) concluded that designers rated the novelty of solution ideas inspired by the more abstract descriptions higher than the descriptions closer to the biological system.

Their flexibility also allows them to be used in many different circumstances – they can be used both for biology push and technology pull bio-inspired design. They are most suitable for creative ideation. A study has shown that teams using *Biocards* generated a higher percentage of novel solution ideas and concepts than teams using brainstorming (Keshwani et al. 2013). But *Biocards* can also be used for the application of biological knowledge and collaborative research projects. In any collaboration, their one-page format and visualization of the functional principle seems beneficial for communicating ideas.

On the downside, the high degree of simplification can also prove to be a disadvantage: Biological phenomena can rarely be reduced to one physical principle. In the case of the torrent frog, it is still possible to depict several physical principles, as we assume that they work in parallel. But for other biological phenomena this is not the case – *Biocards* do not provide support for the formulation and visualization of multi-faceted principles. This also increases the risks of misunderstandings by the designers that use previously prepared *Biocards*. They can become fixated on a simplified explanation of a biological principle - such as *the frog adheres to wet surfaces due to capillary effects* – and possibly neglect the complexity of the biological system.

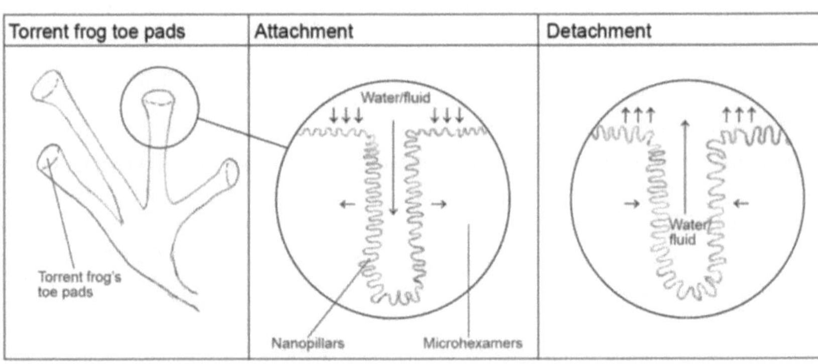

Figure 3-9: Biology-close visualization - capillary effect of torrent frog toes

3.4.2 SBF Modeling

Fact sheet p. 181

SBF modeling is an approach related to the technical abstraction concepts of *components*, *organs,* and *functions* introduced in sub-section 3.1 (p. 55). SBF stands for *structure* (similar to *components*), *behavior* (similar to *organs*), and *function*. SBF models consist of three separate models for structure, behavior, and function of a system. The three parts have been re-named to *What*, *Why* and *How* to avoid misunderstandings regarding the terms *structure*, *behavior* and *function*. SBF models can be equally used to represent technical and biological systems (Goel et al. 2009; Yen et al. 2014).

Structure (*What?*) describes the components of a system and their relations. Function (*Why*?) describes the initial and desired state. Behavior (*How*?) describes a causal pathway to transfer the initial into the desired state (Yen et al. 2014).

The fact sheet shows a SBF model of the medical plaster example modeled according to the instructions given by the Georgia Institute of Technology (2016):

The structure (*What?*) consists of components (elements), substances (flows) and their connections. The medical plaster would have two components (plaster, skin) and one substance (medicine) identified as the main components. Relevant properties and their values are listed in the boxes of each component or substance. It must be noted that the structure is a representation of the system in the initial state (Georgia Institute of Technology 2016). Therefore, in our example, the component *skin* has the property *condition* with the value *wounded*. The connections describe the physical relations between the components and substances. For example, the components *skin* and *plaster* are *attached to* each other. Other connection types indicate that one substance/component is *contained in* or *rests on* another.

The function (*Why?*) describes the purpose of the system: a transformation of a beginning state to an objective state. Regarding the medicine plaster, the beginning state is a *wounded skin*, the objective state a *healed skin*. One must note that *skin* is a component from the structure representation. One or several of the components and substances from the structure representation are re-used for describing the beginning and objective state. The selected components/ substances are the central element of the function representation. There are three types of functions:

- *Accomplishment function*: beginning and objective state of the selected components/substances are different
- *Maintenance function*: beginning and objective state of the selected components/substances are the same, intermediate states can be different
- *Preventative function*: beginning, intermediate and objective state of the selected components/substances are the same

Our example describes an accomplishment function. An example for a maintenance function is an air-conditioning system which continuously measures the room temperature and blows in cold air when the room temperature rises. An example for a preventative function are door locks – the overall goal is that the lock remains closed at any time, it *prevents* the opening of the door

The behavior (*How?*) describes the intermediary states between the beginning and objective state. As the plaster example on the fact sheet shows, it is more detailed than the function representation. It includes the state transitions of all components and substances. Moreover, it explains how the transition occurs. There are several transition types:

- By function:
  The transition occurs due to a function performed by a sub-system (this so-called sub-function can be modelled as a separate SBF model)
- By state:
  The transition is enabled by the achievement of a state of another component (e.g. electricity can move in a technical system when the switch is activated)
- By transition:
  The transition occurs simultaneously due to another transition (e.g. a gear rotates counter-clockwise because of the clockwise rotation of another gear)
- By external stimulus:
  The transition is triggered by an external stimulus (for example the user pushing a button)
- By principle:
  The transition occurs due to a physical law or principle (e.g. gravity)
- By structural connection:
  The transition occurs because a component is connected to another component (e.g. the movement of a screw is determined by the connection between its thread and another components thread)

Often, more than one transition type enables a transition. For example, the rotation of a gear is enabled by the rotation of another gear (transition type: *by transition*) and by the physical connection to the other gear (transition type: *by structural connection*). Sometimes, these obvious transition types are neglected in SBF models.

The transition type *by function* is a central concept for the decomposition of functions. In the medical plaster example, transition 2-3 enables the change of the location of the medicine: in the initial state, medicine is located in the plaster, and in the objective state it is located on the wound. How does this transition occur? The simple SBF model depicted on the fact sheet cannot explain the transition. Instead of enlarging the model, a sub-function can be represented in another SBF model. This sub-system has its own structure, function, and behavior model. A component of such a sub-system could be a porous material patch on the plaster which releases the medicine. By means of decomposition, a hierarchical structure of a system can be built. For the medical plaster, this is shown in Figure 3-10. In addition to the *release medicine* function, the function *adhere to skin* can be decomposed. This involves modeling the glue and its function *build chemical connection* to the user's skin.

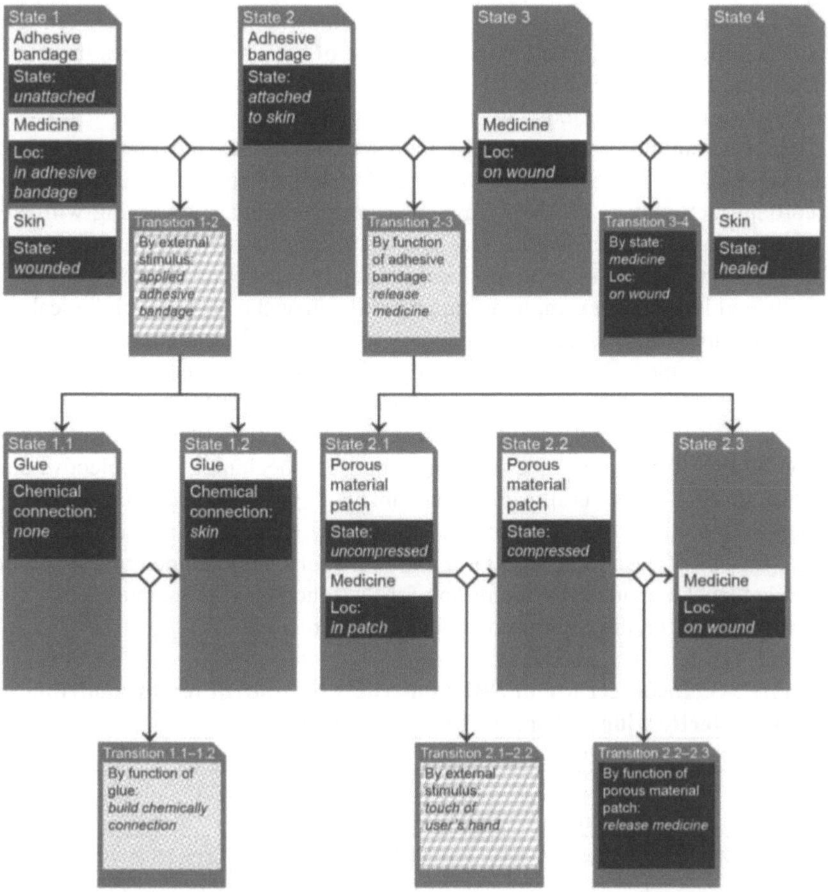

Figure 3-10: Hierarchical structure of SBF models – medical plaster example

This example shows that it is possible to decompose functions on several hierarchical levels which facilitates the understanding of the system. However, there is a limit to its usefulness: in our example, decomposing the function *build chemical connection* would not make much sense for most plaster designers if we assume that they do not develop glues, but rather use glues from other manufacturers.

To summarize, the advantage of SBF models is their high level of detail: it is possible to model any system in a step-by-step approach by starting at a high hierarchical level with a few components and detailing it more and more. As both biological and technical systems can be modelled, SBF modelling is suitable for biology push and technology pull approaches in bio-inspired design.

Moreover, SBF models require very formal and general descriptions (for example naming the physical laws that enable a transition). These formal and general descriptions can support the understanding of details in interdisciplinary collaborations. Therefore, SBF models are suitable for collaborative research projects and for the application of existing biological knowledge.

However, building a detailed SBF model requires time, modeling experience and expert knowledge. SBF models are therefore not suitable for creative ideation. According to Yen et al. (2014), a major obstacle for students working with SBF models in class is the choice of adequate abstraction levels, that is, they tend to jump between abstraction or hierarchical levels. In most cases, students deviate to very general levels, for example discussing survival as the goal of a biological system. Sometimes, they concentrate on overly low levels of abstraction, explaining a phenomenon in more detail than necessary. This also leads to confusion regarding the three separate representations for structure, function, and behavior. For example, students try to model a function on several hierarchical levels instead of focusing on one function and the transitions between beginning and objective state (Yen et al. 2014). This is illustrated in Table 3-3: the observed student team regarded the functions solar absorbance and energy capture of iridescent butterfly wings. Instead of describing the states and transitions, they jumped to a higher hierarchical level and analyzed the reason for the functions – to maintain body temperature. Moreover, they confused behavior and structure.

Table 3-3: Incorrect use of SBF constructs by student teams modeling iridescent butterfly wings (observed by Yen et al. 2014)

SBF constructs	Student team	Correct
What? (structure)	Solar absorbance, energy capture	Structural color patterns
Why? (function)	Maintain body temperature	Solar absorbance, energy capture
How? (behavior)	Structural color patterns	States and transitions

Finally, the hierarchical nature of SBF models supports a top-down approach which is typical for technical product development but not common in biological research. As discussed previously in this book, the focus on functions represents a technical viewpoint and technical terminology. Therefore, even though SBF models can be used to model technical and biological systems, biologists might have more difficulties using them than engineers.

3.4.3 SAPPhIRE

Fact sheet p. 183

Similar to SBF modeling, the SAPPhIRE representation has been developed based on technical abstraction concepts of *components*, *organs,* and *functions* introduced in sub-section 3.1 (p. 55). Unlike SBF modeling, which represents structure (components), function, and behavior (organs) separately, a SAPPhIRE model links the concepts in one model. It aims at representing both biological and technical systems. A SAPPhIRE model consists of seven constructs. It enables a step-by-step abstraction from the component level (construct *parts*) to the function level (construct *action*). The constructs in between describe the behavior of the system or in other words they show *how* the function is achieved (Chakrabarti et al. 2005; Goel et al. 2009).

The fact sheet shows the example of the torrent frog's adhesion to surfaces:

On the component level, ***parts*** describe the physical components and interfaces that constitute the entity of interest and its surroundings. Regarding the torrent frog adhesion, relevant parts are the hexamer microstructure and the pillar-shaped nanostructure. Moreover, fluid can also be considered a component of the system.

The parts create ***organs***, i.e. a set of properties and conditions of the entity and its surroundings required for an interaction between them. In the case of the torrent frog, both hexamers on micro-level and pillars on the nano-level create space for the fluid.

Moreover, additional **inputs** are necessary for the interaction. An input is a physical variable that crosses the boundary of the entity and is essential for an interaction between the entity and its surroundings. For example, the fluid's viscosity plays a role in the interaction between the structure of the torrent frog's toe pads and the fluid.

Together, organs and inputs enable a ***physical effect***, i.e. a principle of nature that underlies and governs an interaction. In our example, the space for the fluid in the structure of the torrent frog's toe pads and the fluid's low viscosity would enable a capillary effect.

The physical effect leads to a ***physical phenomenon***, or in other words, an interaction between the entity and its surroundings. For example, the capillary effect causes adhesion between the frog toes and the surface the frog clings to.

The physical phenomenon enables a specific **state**, this means that either a property of the entity (or its surroundings) that is involved in an interaction remains constant, or that it changes due to the physical phenomenon. Here, normal forces between the toe pads and the surface would be increased by the adhesion.

On the most abstract level, the state or change of state can be interpreted as an ***action***. An action is the abstract description or high-level interpretation of an interaction between the entity and its surroundings. As to the frog example, the described behavior results in the frog's adhesion to surfaces.

In conclusion, SAPPhIRE allows to analyze and represent technical and biological systems in detail. It is therefore suitable for both biology push and technology pull approaches in bio-inspired design. It enables a step-by-step abstraction and can also support the detailed understanding in interdisciplinary collaborations. However, SAPPhIRE has been developed from a technical viewpoint and uses a technical terminology which can be unfamiliar to biologists. Moreover, it requires time, modeling experience and expert knowledge to build a SAPPhIRE model.

Moreover, SAPPhIRE models are sequential representations of usually only one physical principle. This is illustrated by the example used on the fact sheet: the frog's adhesion is probably enabled by several physical principles, friction, Van-der-Waals forces, etc. – the capillary effect is just one of them. To represent the adhesion principles adequately, several SAPPhIRE models must be developed.

3.4.4 KoMBi

Fact sheet p. 185

The *KoMBi* abstraction approach has been developed for a communication platform for bio-inspired design. The aim is to enable both biologists and engineers to represent biological or technical systems using the same constructs. The simplified representations should facilitate an initial interdisciplinary understanding with possible collaboration partners. *KoMBi* is based on an analysis of visual representations in biological textbooks and product development literature. By means of this analysis, six features were identified (Hashemi Farzaneh et al. 2015; Hashemi Farzaneh et al. 2016):

- Morphological feature: representation of the morphology of a biological system, i.e. the shape and/or structural relations of its elements
- Relational feature: representation of cause-and-effect relations between several biological systems or system elements
- Change feature: representation of the change of a biological system or system elements
- Data feature: representation of quantitative or qualitative data acquired on a biological system or its elements
- Mathematical feature: a mathematical representation of a biological system or its elements
- Comparative feature: representation of a comparison between several variations of biological systems or its elements

The features were incorporated in representations for system description (morphological feature) and system behavior (relational and change feature) and properties (data and mathematical feature). The comparative feature can be represented by checking system description and behavior, and properties of several biological or technical systems against each other (Hashemi Farzaneh et al. 2015). The usability of the representations and guidelines was tested with undergraduate and graduate biologists and engineers. Based on the evaluation, the representations were improved and implemented in a software prototype and web application. Independent of the two pieces of software, *KoMBi* representations for system description and behavior and property can be drawn with paper and pen.

The fact sheet shows a *KoMBi* representation of the medical plaster example.

In the first step, the system description must be set up. As the tests have shown that experts tend to go into too much detail (Hashemi Farzaneh et al. 2015), only five system elements should be named for the system description. These elements can be linked by two relation types: *consist of* and *connected to*. In example found on the fact sheet, the element *plaster consists of* the elements *adhesive*, *base,* and *contact material.* It is connected to the element *medicine.*

In the second step, system behavior and properties are modeled. For this step, the system elements from the system description can be re-used, but additional elements can be defined. For the medical plaster example, the elements *adhesive* and *base* were abandoned because their role is not in focus here. The focus of this sample development is the application of medicine by the plaster. For this purpose, only the elements *contact material* and *medicine* play a role. Moreover, the additional element skin is used. How do these elements interact?

This can be modeled using relations and properties: in the example, the location of medicine is described by two properties. Medicine has the property *in contact material* which changes to *on skin.* This change is triggered by an influence-relation from the contact material: contact material *emits* medicine. Similarly, medicine *heals* (influence-relation) the *wounded* (property) skin, so that it changes to *healed* (property).

To summarize, there are three relation types: elements can be related to properties by a *has*-relation (e.g. skin *has* the property wounded/healed). Moreover, they can be related to properties by an *influence*-relation that is explicitly named (e.g. *emit*/*heal*). Properties are related to each other by a *changes-to*-relation (e.g. the wounded *changes to* healed).

Properties can be of very different nature; they can represent a location or a state, but also detailed data such as weight or size. It must be noted that properties do not necessarily have to change as shown in the example – they can also remain constant. For example, the size of the contact material could be defined in the system behavior & properties example, as it is relevant for the amount of medicine to be emitted as well as the size of the wound.

To conclude, *KoMBi* can be used both for biology push and technology pull approaches. In contrast to other approaches, it incorporates both a technical and a biological perspective. It also shares similarities with the SBF modeling approach. To a certain extent, the *KoMBi* system description is similar to the SBF structure representation. *KoMBi* system behavior & properties is comparable to SBF behavior representation. However, *KoMBi* is less formalized and less detailed – this has advantages and disadvantages. *KoMBi* requires less effort and expertise but represents less detailed explanation of the represented systems. Nevertheless, it still requires a certain time-effort and is therefore not useful for creative ideation. It can be used both for the application of biological knowledge and collaborative research projects. To find collaboration partners in a Bio-inspired Open Innovation approach, *KoMBi* representations can be (see fact sheet p.192).

4 Search

Finding suitable biological or technical systems is one of the major challenges of bio-inspired design. Numerous researchers have developed methods and tools to support systematic search for biological or technical systems (e.g. (Baldussu 2012; Vandevenne et al. 2015; Helms 2016).

How can we search effectively without losing ourselves in the face of countless possible biological or technical inspirations? To provide guidelines for effective search, we first discuss a general perspective (4.1). Second, we present intuitive and discursive approaches from technical product development (4.2). Third, we explain search methods that were developed for bio-inspired design. We present sources for biological and technical information in section 4.3 (*where* to search?) and strategies for formulating search queries in section 4.4 (*how* to search?).

4.1 General Perspective on Search

From a general perspective, solution search is linked to the process of problem solving. A problem has an undesired current state that must be transformed into a desired final state. The transformation is complicated by a barrier between the two states (Dörner 1987, p.10).

Dörner (1987, p.10–14) defines three types of barriers, depending on the clarity of the goal and the familiarity of the means to reach it. They are illustrated in Figure 4-1.

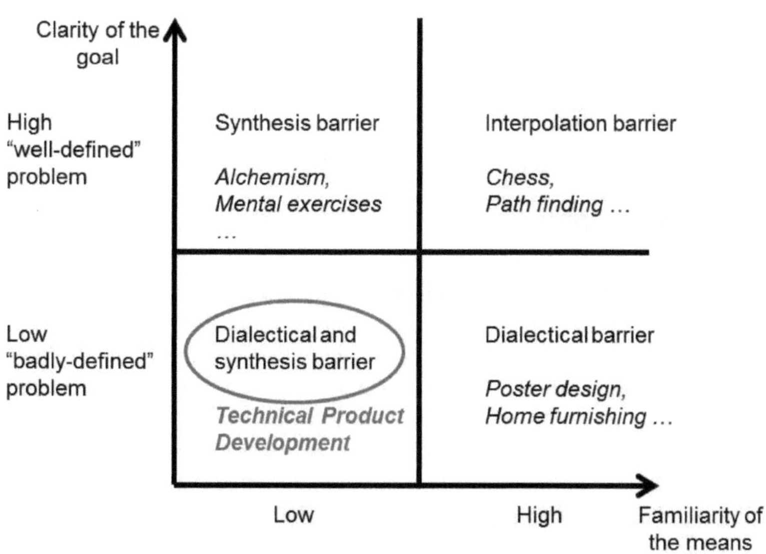

Figure 4-1: Barriers for problem solving (McCarthy 1956; Dörner 1987, p.10–14)

H. Hashemi Farzaneh and U. Lindemann, *A Practical Guide to Bio-inspired Design*, https://doi.org/10.1007/978-3-662-57684-7_4

The first dimension to consider is the clarity of the goal. Goals that are clarified can also be regarded as being well-defined problems (McCarthy 1956). A well-defined problem has a defined, testable final state. For example, the game of chess implies a highly clarified problem: the desired final state is to checkmate the opponent. The second dimension is the familiarity of means to achieve this. In chess, the players know the rules of the game. As such, it is clearly defined how each game piece can be moved.

Consequently, playing chess has an **interpolation** barrier – the players know both the goal and the means to reach it. The difficulty of the interpolation barrier lies in the number of possible combinations of moves. A similar problem is path finding. The desired final location is known, and the possible means to get there too (e.g. walking, by car, by public transport). The difficulty again lies in calculating the quickest way of reaching the desired location by considering different possible combinations of transportation means and paths.

If the goal is clarified, but the means of reaching the goal are not, the problem poses a **synthesis barrier**. This is the case with alchemism for example: alchemists knew that they wanted to synthesize gold, but they did not know which substances or which procedure to use. Many mental exercises also fall into this category. Figure 4-2 shows an example: for the nine-points-problem, the points have to be connected by four lines in one stroke. Synthesis barriers often require *thinking outside the box,* beyond our habits and usual ways of thinking. (If you want to solve the problem, the solution is depicted in Figure 4-4, p. 84)

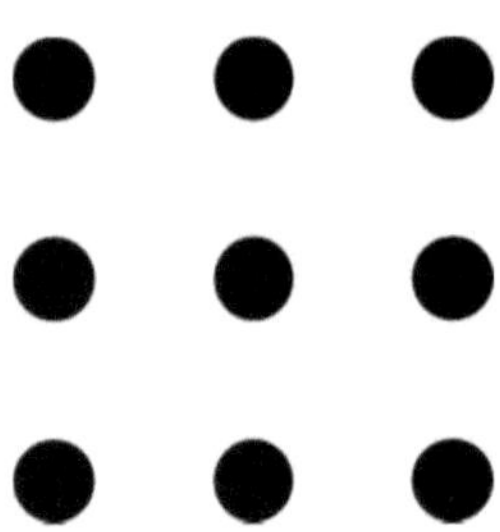

Figure 4-2: Mental exercise "Nine-points-problem" (Dörner 1987, p.77)

In the case of a badly-defined or unclear goal, the problem poses a **dialectical barrier**. An example is the design of a poster. We know the available means, i.e. the poster format, the script, and visual elements that we can use. However, there is no unambiguous definition of the desired goal. What does a good poster look like? How can it best communicate information? There are infinite answers to this depending on the poster designer, the audience, the place where the poster will be installed etc.

Returning to technical problems, they often imply that both the goal and the means to reach it are not clearly defined or are unknown. They therefore pose both a synthesis and a dialectical barrier. In technical product development procedures, the explicit discussion of the task or technical problem aims to overcome the dialectical barrier. By defining the exact requirements, designers clarify the goal of the development process and make it testable (see chapter 2).

Adopting a cognitive perspective, psychologists have identified general problem solving strategies. Anderson (2009, p.209–241) describes the cognitive procedures **difference reduction** and **means-end analysis** illustrated in Figure 4-3.

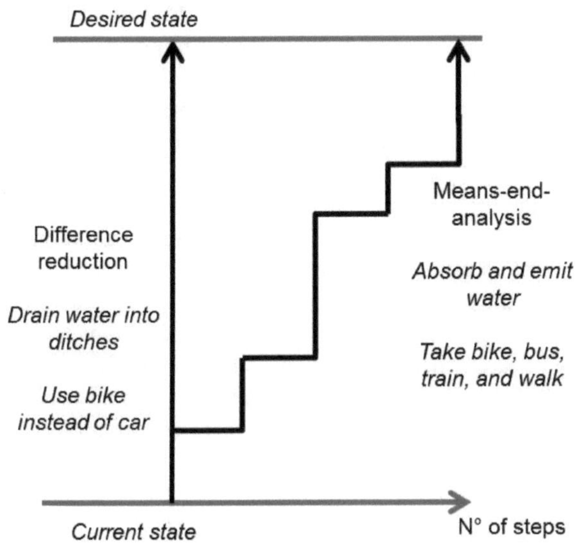

Figure 4-3: Difference reduction and means-end-analysis (based on Anderson 2009, p.209–241)

Difference reduction presumes the identification of the difference between the current state of a situation and the desired state (the problem needs to be well-defined for this!). For example, as with the drainage task, the difference between the current and the desired state is the amount of water contained in the soil. If we want to transport our shopping from a shopping mall to our home, the difference between the current state and the desired state is a given distance in space. After identifying the difference to be reduced, there are different strategies to overcome it. For example, a solution to reduce the wetness of the soil is draining the water into ditches. To overcome a distance, one can use a car, a bike or other means of transportation.

Means-end analysis is used for less obvious solutions: instead of regarding only the overall goal – the desired state – a stepwise approach must be taken. For example, the drainage problem is complicated by the fact that the soil can not only be too wet, but also too dry. These two cases have to be considered for finding an adequate solution. Regarding the transportation task, we might have to use several means of transportation if we do not have a car to overcome a longer distance. For example, we might have to walk to the next bus station, wait for a bus, take the bus, get out at a train station, take a train, and so on.

Dividing an overall goal (or function) into sub-goals (or sub-functions) is a commonly postulated strategy for technical product development (e.g. Pahl and Beitz 2013, p.170–178). The idea is to reduce complexity and to generate partial solutions to each sub-goal separately. Methods support the combination of these sub-solutions to an overall solution (Ponn and Lindemann 2011, p.116–117). On the one hand, this strategy can facilitate in achieving the overall goal, whilst on the other hand it can result in more alternative solutions.

Even if the overall goal is divided into sub-goals, the generation of partial solutions remains a challenge: how can solution ideas be found if the means to reach the goals are unknown?

Looking again at the nine-points-problem in Figure 4-2, it illustrates how individuals tend to search in a search space which is smaller than the solution space: the solution to the nine-points-problem can only be found if one draws a line beyond the area found between the points.

This observation can be transferred to technical product development: designers frequently restrict their search space to familiar solutions that they deem adequate for the technical task. However, the solution space goes beyond these familiar solutions. There are unknown solutions, and then there are solutions that the designers know, but fail to recognize their relevance. In the next sub-section, we show approaches from technical product development to the search of these solutions.

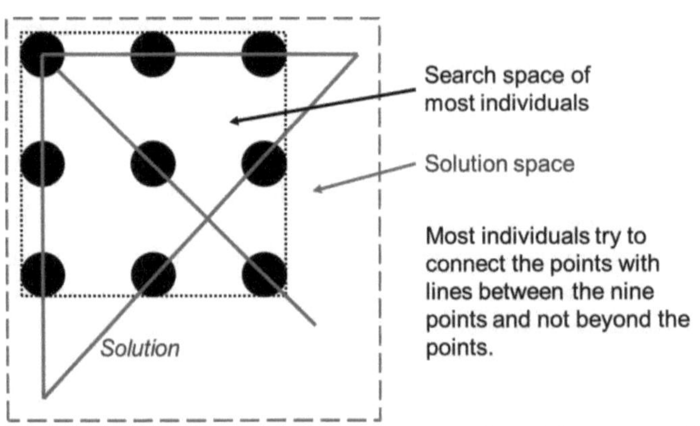

Figure 4-4: Solution to "Nine-points-problem" (Dörner 1987, p.77)

4.2 Search in Technical Product Development

Figure 4-5 illustrates different sources of inspiration and solution ideas for designers. In technical product development, the starting point for solution search is a technical task or defined goal of the development project. For generating solutions, first of all, internal knowledge plays a role, such as memories and experiences with existing products to generate solution ideas. Moreover, designers consciously and unconsciously encounter external information that can inspire solution ideas.

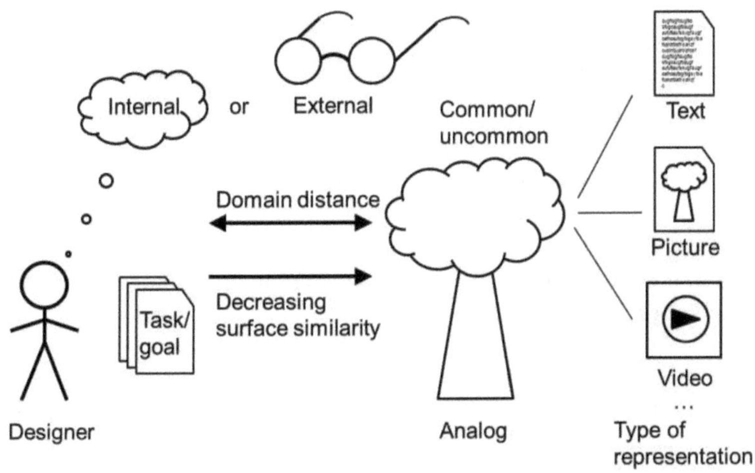

Figure 4-5: Sources of inspiration and solution ideas (Hashemi Farzaneh 2016, p.22)

Internal knowledge and external information can be categorized according to their *domain distance* to the technical task. If the task is to develop a robot gripper, for example, information on existing robot grippers is very "close" to the technical task. Knowledge of other technical products such as vacuum cleaners can be considered more "distant", but it is still in-domain information. Information from the biological domain, such as information on the torrent frog, is very "distant" from the technical task. With increasing domain distance, the surface-similarity of the inspirational sources increase. This also makes searching for the information more difficult.

The usefulness of knowledge and information for ideation can therefore be represented by the curve shown in Figure 4-6– if it is too "close", it cannot provide novel ideas, if it is too "distant", designers cannot relate it to the technical task (Fu et al. 2012; Gonçalves 2016). As depicted in Figure 4-6, biological information and knowledge has a medium or high distance to technical tasks. It is therefore never too "close", but could be too "distant" to be useful as inspiration.

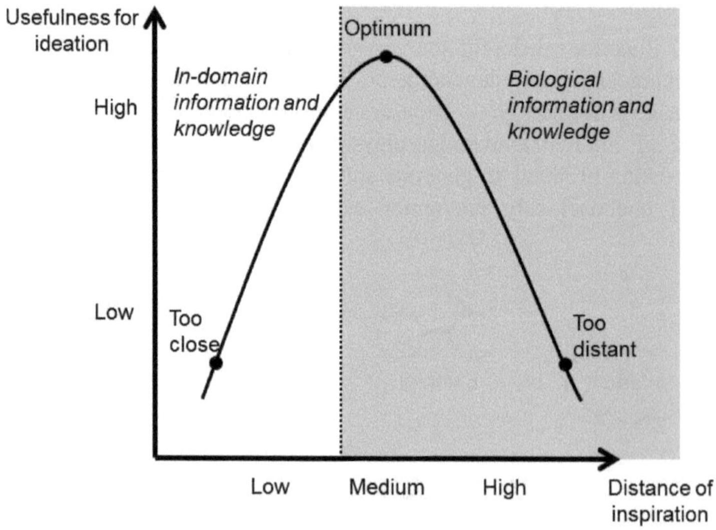

Figure 4-6: Usefulness of information depending on distance from domain (based on Gonçalves 2016, p.144)

Finally, information can be encountered in different representations, such as text, pictures, and videos. The type of representation influences the use of the information for solution ideas. Designers seem to prefer certain types of representation. In a study with student and professional industrial designers, a preference for inspirational images and objects over text was revealed (Gonçalves et al. 2014). Likewise, in simulated unlimited internet search, industrial design students looked for images rather than for text (Gonçalves et al. 2016). This can be explained by the lower cognitive effort that images require (Ware 2012, p.23) – they can be understood in a fraction of the time, whereas a designer needs to read and understand a text. Accordingly, Sarkar and Chakrabarti (2008) observed that designers generated more solution ideas inspired by video and images than by graph representations containing text. However, text also provides more detailed information (Hashemi Farzaneh 2016) and can therefore provide effective inspiration (Goldschmidt and Sever 2011; Gonçalves et al. 2012).

Apart from search on existing products (patent databases, product fairs etc.) and search for inspiration in other domains, two fundamentally different approaches are proposed in technical product development: intuitive and discursive methods (Pahl and Beitz 2013, p.77–102; Lindemann 2016). Intuitive methods rely on the designers' knowledge and experience to generate solution ideas. They are often conducted in groups, so that designers can be inspired by others' knowledge and ideas. Discursive methods push designers to structure the technical task and their ideas. Additional information on existing technical solutions can be provided. Both approaches have been incorporated into search methods for bio-inspired design.

4.2.1 Intuitive or Creativity Methods

Numerous intuitive or creativity methods that have not been developed for a specifically technical context are commonly used in technical product development (Lindemann 2016). Most of them propose group or team work. Nevertheless, research has shown that individual ideation phases can result in more or better ideas than pure group work (Ulrich and Eppinger 2003, p.108). Therefore, several creativity methods incorporate individual and team phases.

Independent of the creativity method used, creative or intuitive ideation incorporates elements of immersion, incubation, and illumination, as shown in Figure 4-7. Immersion can be supported by a number of activities: designers can gather information and analyze the task as well as possible solutions. Moreover, experiments with very simple prototypes also support designers in understanding the task more fundamentally. A similar result can be achieved by activities of playing or idealization of the task. These less serious activities also lead to a relaxed but concentrated atmosphere that can be fruitful for ideation. After becoming immersed in a task, the new information must be processed. Therefore, a period of incubation often precedes the actual generation of useful ideas. The duration of incubation can vary considerably, from seconds to days or even weeks. Illumination and the generation of useful, often unexpected solution ideas tends to come in unexpected moments.

For the success of creative or intuitive methods, the activities enclosing the actual creative process play a crucial role: the situation analysis and goal definition provides important information for solving the technical task. The activities following the actual idea generation are crucial too: even if idea selection, testing, and evaluation are not part of the method, they are always conducted at least implicitly. In this case in particular, there is the risk that ideas become abandoned ad hoc due to cognitive selection mechanisms, so-called heuristics (Gilovich and Griffin 2002; Hashemi Farzaneh et al. 2013). An example is the tendency to look for the confirmation of existing beliefs, i.e. designers will unconsciously prefer solutions that are in accordance with his or her beliefs (Hallihan et al. 2012). To mitigate the effect of heuristics, creativity or intuitive methods often demand the documentation of all generated ideas and often also an explicit testing and evaluation phase.

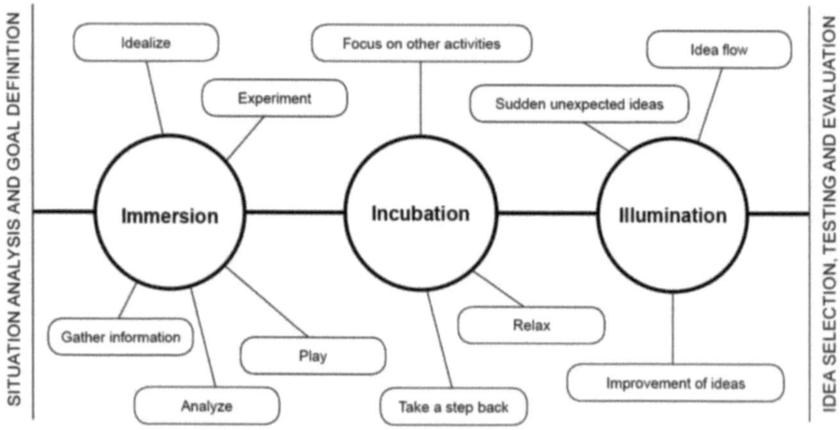

Figure 4-7: Elements of creative or intuitive ideation

The perhaps most common creativity or intuitive method is Brainstorming. Brainstorming is a group creativity method which aims to produce a flow of as many ideas as possible. Brainstorming should be conducted with interdisciplinary groups of 5-15 participants without any hierarchical structure. There are no formal steps, but a couple of rules should be kept in mind (Osborn 1963, p.166–196; Ulrich and Eppinger 2003, p.108; Lindemann 2009, p.142–145; Pahl and Beitz 2013, p.83–85):

- the quantity of ideas is more important than their quality
- ideas must not be criticized
- ideas should be taken on board and further developed
- all ideas should be documented

Brainstorming has been criticized because studies have shown that potential participants generate more ideas individually than in a brainstorming group. This can be explained by the participants' fear to be criticized, the distraction due to others' ideas and loafing in groups. Nevertheless, brainstorming is a popular method in companies. One reason for this is that the participants feel involved in the ideation process and are more likely to accept the outcome (Furnham 2000).

To initiate further ideas in a brainstorming session, there are techniques such as Osborn's questions (Osborn 1963, p.229–290) or the SCAMPER checklist (Serrat 2017). They propose to systematically ask questions to shift the designers' perspective so as to, for example, adapt, magnify, rearrange, or combine the previously generated ideas. Similarly, the TRIZ methodology proposes *thinking in 9 boxes,* which means considering a technical solution's super-system, system, and subsystem in the past, present, and future (Adunka 2016).

Another group technique is the *Six Thinking Hats* method. The participants have to put on different *hats* that symbolize specific perspectives. By using one hat at a time, the participants of the group have to adopt one perspective and think in parallel (Bono 2000, p.4).

The *Synectics* method proposes interdisciplinary collaboration and search. It proposes the formulation of different analogies following a step-by-step approach (Gordon 1961). Using *Synectics* for the transfer of bio-inspired analogies will therefore be presented in detail in section 6.3.6 (p. 163, fact sheet p. 221).

Several creativity or intuitive methods combine group work with individual work:

The *Method 635* proposes that six participants develop and document three ideas individually for a period of five minutes. These documented ideas are then handed to the next participant who has five minutes to further develop it. This is repeated until each participant has worked on each idea. Then, all ideas are reviewed by the group (Lindemann 2009, p.278; Pahl and Beitz 2013, p.85–86)

The *Gallery Method* starts with an individual ideation period in which the participants sketch ideas. Then these ideas are presented in a *gallery* and discussed in the group. Based on the discussions, the participants individually develop the generated ideas and generate new ones. In the selection phase, the group evaluates the developed ideas (Ulrich and Eppinger 2003, p.110; Pahl and Beitz 2013, p.86).

The *Delphi method* consists entirely of individual ideation, i.e. it proposes expert surveys: experts are asked for their opinion on a topic or for solution proposals. Their answers are compared and the experts are asked to comment on the differences. This can be repeated several times. A disadvantage of the method is its long duration. Additionally, the selection of experts can be challenging (Lindemann 2009, p.79; Pahl and Beitz 2013, p.87).

Table 4-1: Intuitive or creativity methods for search used in technical product development

Group work	Group and individual work	Individual work
Brainstorming	Method 635	Delphi method
Osborn's questions, SCAMPER checklist	Gallery method	
Thinking in 9 boxes (TRIZ)		
Six thinking hats		
Synectics		

4.2.2 Discursive or Structured Methods

Discursive or structured methods urge designers to structure and to abstract their technical task. Based on the structure and abstraction, information on existing technical solutions and general physical relations can be accessed. This approach is supposed to stimulate creativity by giving designers a starting point for ideation. Discursive or structured methods can be combined with intuitive methods (Pahl and Beitz 2013, p.89).

To structure a technical task, the methods presented in section 3 may be used. In particular, flow oriented modeling (section 3.1.1, page 59) supports the formulation of the technical task in highly abstract functions. Design catalogs (e.g. by Roth 2000; Koller 2011) list existing technical solutions and general physical relations for abstract functions. Example entries of design catalogs are shown in Figure 4-8. To enlarge or reduce a length, a diameter or a volume, the catalogue suggests using for example a lever, a wedge, or capillary tubes. The catalogue shows a sketch and the equations for calculating the desired effects. The example shows that the suggested solutions are only applicable under specific conditions. The capillary effect, for example, is only useful if the tubes have very small diameters (however, water can then be transported to the crown of the highest trees!). A limitation of design catalogs is that they lack information on the conditions that have to be fulfilled for using the proposed solutions. Nevertheless, catalogs can provide a starting point for solution search. They can also help in overcoming mental blocks when designers run out of ideas during creative ideation.

Physical effect	Physical law	Application example
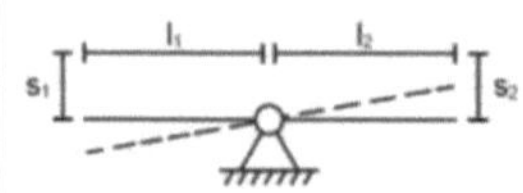	$s_2 = s_1 \cdot \frac{l_2}{l_1}$	Lever gear Cog wheel
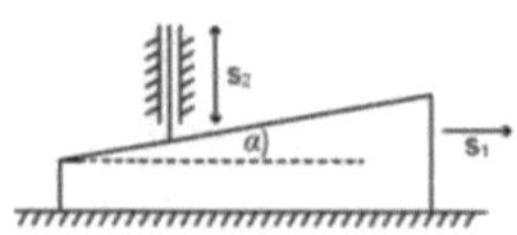	$s_2 = s_1 \cdot \tan\alpha$	Screw Cam mechanism
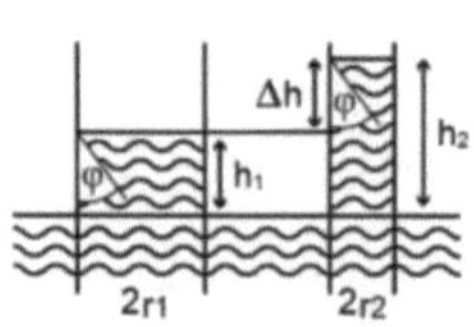	$\Delta h = -\frac{\Delta r}{r_1^2 - r_1 \cdot \Delta r} \cdot \frac{2 \cdot \sigma \cdot \cos\varphi}{\rho \cdot g}$ σ: Surface tension, φ: Contact angle ρ: Density of the fluid	

Figure 4-8: Example for design catalog entries (based on Koller 2011)

For bio-inspired design, the design catalog approach has been used to list biological solutions for abstract technical functions (e.g. by Hill 1997; Gramann 2004). More details on this will be given in the following sub-sections (see section 4.3.3, page 99 and fact sheet page 191).

A different approach to systematically access existing technical solutions is proposed by the TRIZ methodology. To develop TRIZ, Altshuller (1984) analyzed thousands of patents and identified 40 innovation principles. These innovation principles are more general than the physical effects shown in Figure 4-8. One innovation principle is for example asymmetry (principle n°4) (Adunka 2016). An example for using these principles are automotive headlights: the right headlight is directed to the front whereas the left headlight is directed a few degrees to the right side to avoid blinding the oncoming traffic. But which innovation principle is applicable for solving a specific technical task?

As explained in chapter 2 for planning a bio-inspired design project, the *TRIZ-based Procedure* supports the formulation of a paradox (see *TRIZ-based Procedure*, section 2.4.4, p. 44 and fact sheet p. 174)The procedure presented in this book suggests to search for bio-inspired solutions based on the paradox. However, the original TRIZ methodology provides a schema that assigns technical paradoxes to the 40 innovation principles. Technical paradoxes (two conflicting parameters) can be further abstracted to physical paradoxes (one parameter). For physical paradoxes, TRIZ proposes only four general solution strategies (separation of space, of time, by condition, and by system transition) (Adunka 2016).

Based on catalogs, TRIZ, or also creative methods, initial solution ideas are generated. They often do not address the overall goal of the product development process, but rather only partial functions. Classification schemes support the designer to systematically vary the initial solution ideas and find additional solutions. Classification schemes mostly consist of two dimensions. For example, when coating the backside of a carpet, the carpet can be moved as well as the coating applicator. Possible movements are translation, rotation, oscillation, and a combination of several movements. A classification scheme can help designers to find all possible movement combinations (e.g. carpet translates, applicator rotates; carpet rotates, applicator oscillates etc.). A specific type of classification scheme is the morphological box as shown in Figure 2-12 (page 50). The morphological box lists different solutions for each partial function of a technical system. It can also be used to combine different partial solutions with overall solutions.

4.3 Where to Search in Bio-inspired Design

In the previous chapters, it has become apparent that the majority of bio-inspired design methods are based on technical product development methods. This is also the case for search methods – an overview is shown in Figure 4-9. Information sources for bio-inspired design can be clustered similar to information sources for general technical product development: externally, designers can seek for original information. Moreover, based on discursive or structured approaches from technical product development (see section 4.2.2), researchers have developed repositories of prepared bio-inspired design information. Internally, intuitive or creativity methods (see section 4.2.1) can be used to tap the knowledge of engineers and biologists. Studies indicate that this approach is more successful in collaborations between engineers and biologists (Vattam et al. 2008, 2009, 2010; Helten et al. 2011; Hashemi Farzaneh 2016).

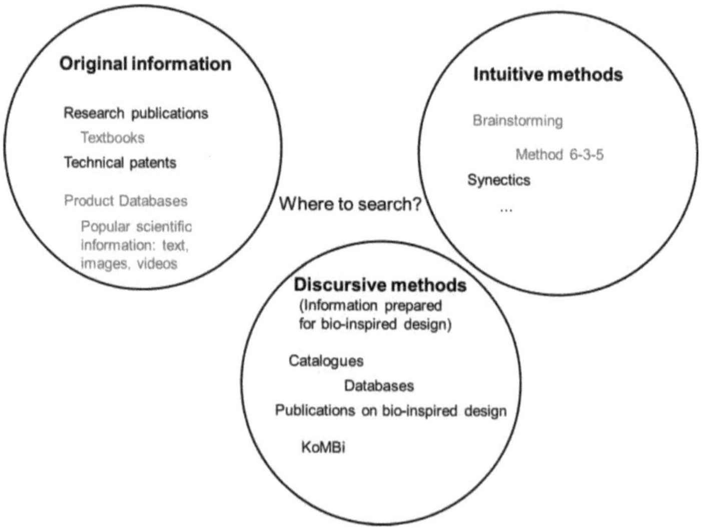

Figure 4-9: Overview on information (and knowledge) sources for bio-inspired design (the methods printed in black are presented on fact sheets)

Most bio-inspired design methods focus on the search for biological systems, i.e. they assume a technical starting point (technology pull approach). Still, search is an important activity for biology push approaches too, as technical application areas must be researched and selected. As no prepared information on technical systems has been assembled for bio-inspired design so far, designers are restricted to original information for this approach.

Original information on biological and technical systems can be found in research publications, such as journal articles, conference contributions, and so on. Details on this approach are discussed in section 4.3.1 and are summarized on the fact sheet on page 187). Textbooks can also be considered research publications. However, especially if they are tailored for undergraduate students, the information is probably easier to understand for designers without disciplinary background. On the downside, the up-to-datedness of the information is probably not as good as in journal articles or conference contributions. Popular scientific information addresses the general public and therefore does not use or explains domain-specific terminology. Moreover, it usually contains illustrating images in addition to text or even videos. This makes them easier to understand for laypeople. However, the simplification of information in popular scientific text, images, and videos may result in facts relevant for the biological or technical system's functioning being omitted. Therefore, popular scientific information can only be used as a starting point in a bio-inspired design project. Furthermore detailed information must be used to understand the biological or technical systems.

As for technical systems, they are provided with information by technical patents, and they can be searched for in patent databases. One difficulty of patents is that they have not been formulated to disseminate information (in contrast to research publications). This poses a considerable challenge for understanding, even for engineers. Details on patent databases are given in section 4.3.2 and on the factsheet on page 189. An additional option are product databases such as thomasnet.com (Thomas Publishing Company 2017). They, however, support searching for products by product names rather than by function or other properties. This would make little sense in the case of a biology push project where biological functions, properties and solution principles are the starting point.

To facilitate the access to information on biological systems, researchers have developed a number of tools. These tools are based on discursive methods for solution search (see section 4.2.2). They provide **information which has been prepared for bio-inspired design**. The advantage is that they are tailored to engineers. However, their disadvantage is that all prepared information is information which has already been used for bio-inspired design. Entirely new inspiration cannot be found (Vincent et al. 2006; Helms 2016, p.26). No similar tools have been developed yet for biologists searching for technical application areas. However, the information can also be used in a different direction. That is, by considering in which areas bio-inspired designs have been used already, ideas for technical application areas can emerge.

To start with, a number of books and other publications such as conference proceedings present examples of biological systems or bio-inspired designs. Examples are given in the appendix (section 9.1, p. 240). For specific broad topics, for example surface design, designers can search the indicated books and other publications. As the collections are not categorized into concrete technical tasks or problems, this search can be rather time-consuming.

To enable a more effective search, catalogs for bio-inspired design have been built (e.g. Hill 1997; Gramann 2004). They present biological systems categorized by technical functions. With increasing digitalization, databases have been developed (e.g. Löffler 2009; Deldin and Schuknecht 2014). The use of databases is discussed in detail in section 4.3.3 and summarized on a fact sheet (page 191).

A different approach has been taken by researchers attempting to fit bio-inspired design into the TRIZ methodology. Examples for TRIZ-based search tools are shown in Table 4-2. One difficulty that can arise when using these tools is that only few application examples have been published and could be used as models for bio-inspired development projects.

Table 4-2: TRIZ

Reference	TRIZ-based Search Tools
Hill 1997	Catalogue of biological systems assigned to functions, integration of the catalogue into *TRIZ-based Procedure* (see fact sheet p.174)
Vincent et al. 2006	Bio-inspired conflict matrix: bio-inspired (instead of conventional) assignment of technical conflicts to the 40 innovative principles
Bogatyrev and Bogatyreva 2014	Bio-inspired rules for eco-design

In the following sub-sections, the search for research publications (biological and technical) and technical patents is discussed. They provide the most up-to-date information, but may be challenging for disciplinary laypersons to understand. Moreover, the use of catalogs and databases of biological principles is presented. These tools can provide a quicker and easier access to biological information even though biological systems that have not been considered for bio-inspired design probably cannot be found.

The suitability of the different methods for bio-inspired design scenarios and approaches is shown in Figure 4-10. Catalogs/Databases of Biological Principles address particularly creative ideation and a technology pull approach, whereas the other methods are more adequate for the application of biological knowledge and collaborative research projects. Technical patents are only useful for a biology push approach.

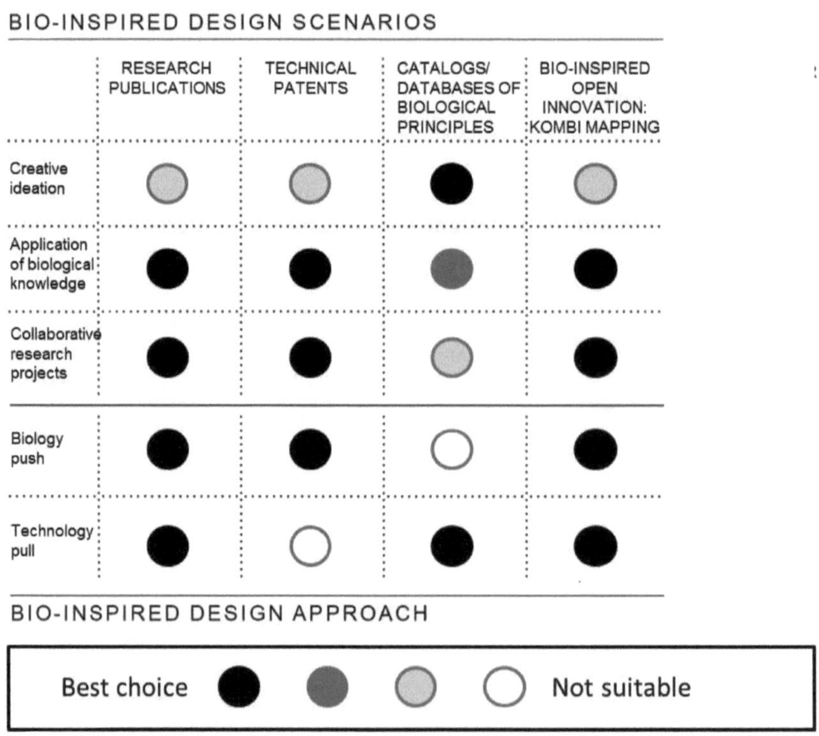

Figure 4-10: Suitability of search methods for different bio-inspired design scenarios and approaches (where to search?)

BEST PRACTICE

Mountain Bike *BIONICON rEVO* – Solutions through Observation in Nature

Bionicon GmbH

Bionicon rEVO,photo: Bionicon GmbH

Mountain bikers face different challenges uphill and downhill: uphill, their center of gravity shifts backwards with increasing slope steepness. Downhill, the center of gravity shifts forwards. This limits the maximum angle that mountain bikers can cycle uphill or downhill. To push these limits, conventional mountain bikes are optimized for better downhill or uphill performance. But why can the bike not adapt to the slope angle, enabling good uphill and downhill performance? One evening in the mountains, the Bionicon team observed a biological species which solves this challenge perfectly: the chamois climbs mountains uphill and downhill keeping its back in a horizontal position by stretching or retracting its legs. Inspired by this observation, the Bionicon team experimented with different technical components to implement the adaption of the bike's front fork. The final solution is pneumatic –air cylinders in the front fork and the bike frame extend or retract. The biker pushes and holds a button which opens a valve to activate the system. To adjust the geometry the rider shifts his weight either forwards or backwards to move the air inside the capsuled system from front fork to rear cylinder and back.

Based on an interview with René Schulz, Bionicon GmbH (July 2017), Bionicon GmbH (2018)

4.3.1 Research Publications

Fact sheet p. 187

The use of research publications was originally proposed for technology pull approaches (e.g. by Kaiser et al. 2012, Vandevenne et al. 2011). Information on the latest research on biological systems can be taken from biological publications. Similarly, technical publications can be used for biology push approaches – information on technical challenges and research on technologies to solve them is presented in research publications from the engineering sciences.

The advantage of research publications is their up-to-datedness. Using research publications, designers can access information that has not yet been considered for bio-inspired design. The level of detail is high as simplifications like those found in popular scientific texts are not made in research publications. Moreover, research publications provide a very broad search space, especially in comparison to catalogs or databases of information prepared for bio-inspired design.

The use of research publications poses several challenges: to start with, finding publications on adequate biological or technical systems can be hindered by domain-specific terminology. In the next section (4.3.4), we discuss this in detail and present strategies to formulate search queries effectively. Recommended search engines either search the entire internet (e.g. scholar.google.com) or only selected publication databases (e.g. pubmed.gov, scopus.com). Search engines provide thousands of results for most search queries. For non-experts in the domain, it is particularly challenging to choose the research publications that are useful. This goes hand in hand with the challenge of understanding and evaluating the suitability of the biological or technical system. Research publications usually address experts in a particular research field and not laypersons in the discipline of biology or engineering. An additional aspect is that research publications mainly present results of specific studies (such as experiments, simulations, etc.). The authors mostly cannot and do not draw conclusions that go beyond these results. Therefore, research publications lack explanations and assumptions which are essential for bio-inspired design. Analyzing the use of visualizations and figures in biological publications, we did not find the explanatory models shown in textbooks for biology students (Hashemi Farzaneh et al. 2016). An example for the use of figures is Endlein et al.; Endlein et al. 2013a (2013b) who compare the adhesion abilities of tree and torrent frogs. They find that torrent frogs adhere better to a wet glass plate with high inclination. They observe that these frogs press their body more effectively on the plate so that the contact area increases. However, there is no figure of this behavior. Instead, they show boxplots of the measured angles of fall as shown in Figure 4-11. The lack of these explanatory figures make understanding the frog's strategies for adhesion difficult for anyone who has no background knowledge and looks at this biological system for the first time.

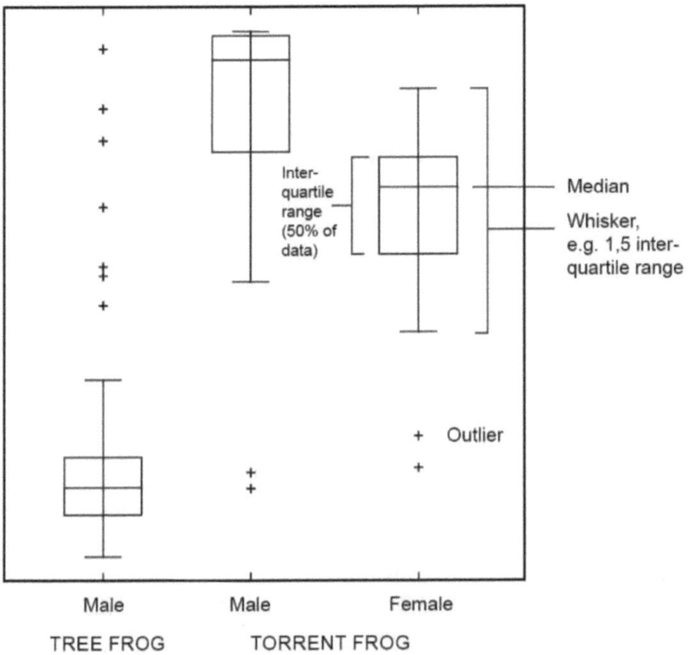

Figure 4-11: Boxplots illustrating the results of adhesion experiments with tree and torrent frogs (modified and labeled from Endlein et al. 2013b)

To tackle the challenges and efficiently use research publications, a reading strategy is recommendable. The list of results displayed by search engines allows to read the title and the abstract of the found research publications. If title and abstract seem relevant, designers can access the publication and read the conclusion. The conclusion summarizes the relevant results. If these are interesting, the figures can be considered. The result section then provides details on the research. If the results give hint towards an interesting biological or technical system, the publication usually must be read from beginning to end. In most cases it will also be necessary to search for further information and publications to understand the information. Search engines facilitate this by automatically displaying the citations of a publication, as well as the publications that later cited the regarded publication.

To conclude, research publications can provide detailed, new information on biological and technical systems (technology pull and biology push approaches). However, the efforts to find, select, and understand the information is considerable. Therefore, they are most suitable for collaborative research projects. In particular, they can be used to gain information on the work of possible collaboration partners before the project starts. Research publications also provide contact details of the authors. Additionally, they may also be used for the application of biological knowledge in bio-inspired design projects with a relatively long timeframe.

4.3.2 Technical Patents

Fact sheet p. 189

For biology push approaches, technical patents can be sought in order to find possible technical application areas for biological solutions. Similar to research publications, technical patents can show new developments which can provide possibilities for bio-inspired design projects. Figure 4-12 exemplifies the first ten results displayed for the search query *anti-slip surface* (ordered by date of submission). We clustered the results into broad categories: three patents are related to handling or assembly in manufacturing processes – a wafer support, a positioning device, and a wrench-type tool. Two other patents describe an exercise mat or cover for exercise devices. Another two patents aim at an ergonomic and slip free design of consumer products – a shampoo bottle, and a paper cup. The remaining three patents do not fall under these categories.

As this example illustrates, patent search can be used at different points of a biology push development project: scanning a number of patents and grouping them into categories can provide initial ideas for technical application areas. If a concrete technical task has been defined, patent search can reveal existing technical solutions that address this task. As in conventional technical product development, this can assist in avoiding problems due to similarities to already patented solutions.

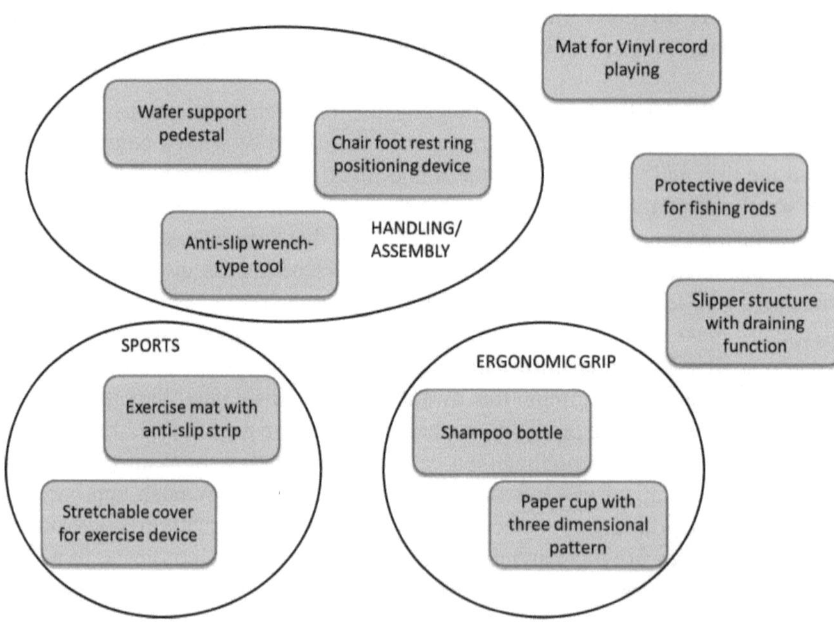

Figure 4-12: Clustering of the first ten search results found by the search query "anti-slip surface" (worldwide.espacenet.com)

Technical patents pose similar challenges to research publications. Finding the adequate technical applications that correspond to the biological solution is hindered by differences in terminology. If the search queries are broad, patent search engines will usually provide a high number of results. Understanding a patent description can prove to be difficult, even for engineers. The reason for this is that technical patents are not formulated to dissipate information (as research publications). Their aim is to formulate a technical solution which has not been published or patented previously. Patent descriptions are therefore simultaneously in some aspects very specific (to meet the gap), and in others quite general (so as to cover as much as possible).

Due to the considerable effort, patent search is less suitable for creative ideation and rather for the application of biological knowledge or research collaborations. To formulate search queries adequately, the *BIOscrabble* approach proves useful (see section 4.4.3 and fact sheet p. 189).

4.3.3 Catalogs/Databases of Biological Principles

Fact sheet p. 191

As discussed in section 4.2.2, catalogs and databases of technical solutions have been developed to support technical solution search. Likewise, catalogs and databases of biological systems have been constructed for bio-inspired design. The first of these approaches were paper-based and contained only a few hundred biological systems. Examples are the *Catalogue of basic functions* and the *Association list* (Gramann 2004). More recent attempts resulted in digital databases, such as the *Database of biomimetic effects* (Löffler 2009). The most well-known, freely-accessible online database with on-going contributions is *asknature.org*. The database provides information about approximately 1800 biological systems (Deldin and Schuknecht 2014). There are several other digital databases such as *Idea-Inspire* (Chakrabarti 2014) or *DANE* (Vattam et al. 2011), but their online access is restricted.

As the table on the factsheet (p. 191) shows, both catalogs and databases provide access to the biological systems via functions. However, the used functions and their level of concretization differ considerably. Table 4-3 shows the functions used by the catalogs assigned to groups of the database *asknature.org*. On *asknature.org*, functions are grouped into broader categories, i.e. subgroups and groups. Subgroups and functions are shown in Table 4-4.

Table 4-3: Functions to search for biological systems in catalogs assigned to *asknature.org groups* (The Biomimicry 3.8 Institute 2016)
Sample entries, see p. 248

Groups	Association list (Gramann 2004)	Database of biomimetic effects (Löffler 2009)	Catalogue of basic functions (Hill 1997)
Move or stay put	Move (vibrate, rotate, lift, move), adjust	Locomotion principles, flow principles, air exchange, piercing- and penetrating apparatus, joining principles	Transfer of material, connect from material (tack, staple)
Protect from physical harm	Maintain	Protection systems, shock absorption, packaging bionic, heat regulation	Support / carry materials, connect from materials (interconnect)
Modify	Modify, deform (smooth, forming), change aggregate state	Explosion mechanisms	Disconnect materials (open from the inside, tear open/ brake up), connect from material (open & close), connect/disconnect energy, form material/energy, support/ carry energy, change material & energy
Make	Connect, produce	Bio-inspired architecture	Connect from material
Process information	Measure/ recognize, partly produce	Sensors, light transmission	Change information, store information, support/carry information, compare information, block information, connect/disconnect information, Form information
Break down	Disconnect	Cutting up apparatus	Disconnect materials
Get, store, or distribute resources	Collect (deposit, absorb, embed, store), extract, partly disconnect, distribute	Storage systems, distribution systems	Transfer of material (draw), transfer of energy, store material (store, reduce water evaporation), lock material, store/lock energy

Note: The databases do not contain functions which can be assigned to the *asknature* group *maintain community*

Table 4-4: Asknature groups, subgroups and functions (The Biomimicry 3.8 Institute 2016)

Groups	Subgroups	Functions
Move or stay put	Attach	Permanently, Temporarily
	Move	In/on solids, In/on liquids, In/through gases
Protect from physical harm	Protect from living threats	Animals, Plants, Fungi, Microbes
	Protect from non-living threats	Excess liquids, Wind, Loss of liquids, Gases, Loss of gases, Dirt/solids, Light, Chemicals, Fire, Ice, Temperature, Nuclear radiation
	Manage structural forces	Shear, Thermal shock, Impact, Tension Turbulence, Mechanical wear, Chemical wear, Creep, Compression
	Regulate physiol. processes	Cellular processes, Maintain homeostasis, Reproduction or growth
	Prevent structural failure	Buckling, Deformation, Fatigue, Melting, Fracture/rupture
Maintain community	Coordinate	Coordinate by self-organization, Activities-Systems
	Cooperate	Within the same species, Cooperate/compete between different species, Within a (eco)systems
	Provide ecosystem services	Manage disturbance in a community, Regulate hydrological flows, Pollinate, Generate soil/renew fertility, Detoxification/purification of air/water/waste, Control erosion and sediment, Regulate water storage, Cycle nutrients, Regulate atmospheric composition, Regulate climate, Disperse seeds, Maintain biodiversity, Biological control of population, pests, diseases
Modify	Modify physical state	Size/shape/mass/volume, Pressure, Density, Phase, Buoyancy, Light/color, Material characteristics, Speed, Position
	Modify chemical/electrical state	Energy state, Free radical reactivity, Concentration, Chemical potential, Reactivity with water, Oxidation state, Electrical charge, Conductivity, Surface tension, pH - Solubility, Electron transport, Chemically generate flow of electrons
	Adapt/optimize	Adapt genotype, Adapt phenotype, Coevolve, Adapt behaviors, Optimize space/materials
	Transform/convert energy	Electrical energy, Magnetic energy, Chemical energy, Mechanical energy, Thermal energy, Radiant energy (light)
Make	Reproduce	Self-replicate
	Physically assemble	Structure
	Chemically assemble	Polymers, Metal-based compounds, Molecular devices, Specific stereoisomers, Mineral crystals, Inorganic com-

Groups	Subgroups	Functions
		pounds, Organic compounds, Attach a functional group, Detach a functional group, Catalyze making of bounds, On demand
Process infor-mation	Navigate	Through air, Through liquid, Over land, Through solids
	Send signals	Light(visible spectrum), Light(non-visible spectrum), Sound, Tacticle, Chemical(odor, taste, etc), Vibratory, Eletrical/magnetic
	Process signals	Differentiate signal from noise, Transduce/convert signals, Respond to signals
	Sense signals/ environmental cues	Light(visible spectrum), Light(non-visible spectrum), Electricity/magnetism, Disease, Touch and mechanical forces, Chemicals (odor, taste etc.), Atmospheric conditions, Sound and other vibrations, Temperature, Motion, Paim, Body awareness, Balance/orientation, Shape and pattern, Time and day length
	Compute, learn, (en/de)code	
Break down	Chemically break down	Cleave heavy metals from organic compounds, Cleave halogens from organic compounds, Inorganic compounds, Polymers, Organic compounds, Catalyze breaking of bonds
	Physically break down	Non-living materials, Living materials
Get, store, or distrib-ute re-sources	Capture, ab-sorb or filter	Organisms, Solids, Liquids, Gases, Energy, Chemical entities
	Store	Solids, Liquids, Gases, Energy, Chemical entities
	Distribute	Solids, Liquids, Gases, Energy
	Expel	Solids, Liquids, Gases

It can be observed that some functions are similarly used throughout the catalogs and databases, e.g. *break down chemically* or *physically* (asknature.org), *disconnect* (association list), *cutting up apparatus* (database of biomimetic effects), *disconnect materials* (catalogue of basic functions). One asknature.org group is not addressed by the other presented approaches (*maintain community*). Other functions focus on different aspects, e.g. *modify* (asknature.org) and *explosion mechanisms* (database of biomimetic effects).
In the appendix (section 9.2, p. 248), some examples which were taken from different sources are listed for each asknature.org subgroup to give an idea of the biological systems suggested by catalogs and databases.

Catalogs and databases give similar information on the biological systems as the purely technical catalogs do for existing technical solutions (see Figure 4-8): they describe the biological system's functional principle in a simplified manner and sometimes provide sketches/photos and give examples on technical products that have been inspired by the biological system.

The advantage of using catalogs and databases is that they provide quick access to biological systems. Expert biological knowledge is not required, since the catalogs and databases use technical terminology. A disadvantage of catalogs and databases is that they only provide information on biological systems which have already been considered for bio-inspired design. Designers cannot expect to find the latest biological research results as with biological research publications. Moreover, catalogs can only be searched by using the functions that they provide. The two discussed databases (database of biomimetic effects, askanture.org) can be searched with any search term. This increases the probability of obtaining results. It does, however, also increase the number of possibly unsuitable results.

As there are no catalogs of technical systems that may be accessed with biological terminology, the approach is limited to technology pull bio-inspired design. As not much time-effort is needed, using databases or catalogs is most suitable for creative ideation. For the application of biological knowledge in more detail and collaborative research projects, the search in catalogs or databases can provide a starting point. As catalogs and databases do not provide very detailed information, they often disregard the fact that biological systems are multi-faceted and apply more than one functional principle. (A good example for this is the adhesion of the torrent frog). Therefore, additional information sources must be considered. As functions provide the best search terms for catalogs and databases, functional modeling is a good method for abstraction to prepare the search.

4.3.4 Bio-inspired Open Innovation: KoMBi Mapping

Fact sheet p. 192

KoMBi Mapping is an approach for Bio-inspired Open Innovation. Open Innovation is a relatively novel approach to technical product development. It includes external in addition to internal partners to develop product ideas and technical solutions. In particular it integrates external knowledge through internet and social media technology (Gürtler 2016, p.2).

In bio-inspired design, direct contact between biologists and engineers can be very useful: if we link them in the search phase, biologists can provide expert information on the biological system, and engineers on the technical system. This expert information can greatly facilitate analysis and comparison of technical and biological systems and transfer of bio-inspired analogies. Open innovation approaches, as well as internet and social media technology provide a chance for establishing contacts between biologists and engineers. These contacts could ideally result in collaborative research projects. Nevertheless, the different terminology and different viewpoints of the disciplines are barriers for linking experts from both disciplines. This is where *KoMBi* comes into play: on the *KoMBi* platform, biologists can model biological systems, whilst engineers can model technical systems or tasks. (For details on the *KoMBi* abstraction see section 3.4.4, p. 78, fact sheet p. 185)

Biological and technical systems are mapped by means of identical and similar elements of the *KoMBi* system behavior and property model. The fact sheet shows the example of the torrent frog's toe pads (biological system) and a medical plaster (technical system). As far as the two system behavior and properties models are concerned, they both have the element *emit*. In addition to this shared element, similar elements can be identified: An example would be to map the elements *on surfaces* (torrent frog toe pads) and *on skin* (medical plaster). The words *surface* and *skin* have a semantic relation. Therefore, they can be related using taxonomies such as the Wordnet® database of the English language. It clusters words based on their semantic and lexical relations (Fellbaum 1999; Princeton University 2010). The taxonomies presented in section 4.4.1 (p. 108) can also be used for mapping biological and technical terms.

By this means, the *KoMBi* platform suggests suitable technical systems to biologists searching for a technical application area for a biological solution (biology push approach). Accordingly, the *KoMBi* platform suggests biological systems to engineers searching for a biological inspiration to solve a technical task or problem (technology pull approach). Along with the suggestion of a technical or biological system, an expert will also come into contact with it. It is possible to contact the experts like in common social media systems and to ask questions.

To summarize, *KoMBi* mapping introduces a novel way of searching for biological or technical systems and potential collaboration partners. On a social media platform, it automates the matching of biological and technical systems and potential collaboration partners. Moreover, *KoMBi* mapping facilitates a first understanding of biological and technical systems for non-experts from other disciplines. *KoMBi* models therefore represent a unique base for discussion between biologists and engineers. Nevertheless, modeling a biological or technical system with the *KoMBi* approach requires time and effort. Therefore, *KoMBi* mapping is more suitable for the first steps of collaborative research projects and application of biological knowledge and less for creative ideation. Moreover, *KoMBi* models are abstract simplifications of biological and technical systems. As such, they can lead to misunderstandings – for transferring concrete analogies, more information is needed than the model provides. *KoMBi* mapping is suitable for biology push and technology pull approaches and must be used in combination with the *KoMBi* method for abstraction.

4.4 How to Search in Bio-inspired Design

A major challenge of searching for biological or technical inspiration is formulating the right search terms. We have all experienced this in our everyday life: nowadays it is easy to find information on almost anything using internet search engines, as long as we know the correct term or name of what we are searching for. Searching becomes much more difficult if we do not know this. If, for example, in our apartment one of the window shutters stopped working because the belt winder is broken, we can find a new one easily if we know the term *belt winder*. If not, we first have to find out what this wheel in the opening next to the window is called. For example, we could search for all products related to window shutters. That often takes a couple of minutes. In bio-inspired design, the challenge of finding an adequate search term is even bigger, because we are facing a dialectical barrier (unclear goal) and a synthesis barrier (unclear how to reach the goal). Instead of having a broken shutter this would mean that shutters do not exist. Our problem would be that we want to darken the room in our apartment. Instead of searching for a belt winder, we now have to find a way of darkening rooms. This opens up infinite ways of conducting searches.

Moreover, a particular challenge in bio-inspired design is the different terminology or domain-specific language of the disciplines biology and technical product development. This often makes it necessary to try out several search terms and combination of terms. This is well illustrated by a search for biological systems reported by Lenau et al. (2011) which is summarized in Table 4-5: in a technology pull approach, an engineering-design student had to solve the technical task *reduce the consequences of car collision*. He was supported by a biological librarian. Several initial search terms were formulated from a mainly technical perspective. These search terms are mostly functional, e.g. *shock absorption* or *control deceleration*. Brainstorming resulted in additional search terms which incorporate more solution-oriented biological aspects such as *shell*, or *protection layer*. An initial search in a library and in the database of asknature.org lead to partly unsuitable results, for example protection in biology is often related to protection from predators. Therefore new search term combinations were built (*protection AND impact*). Moreover, some of the search terms were synonymized. The search then resulted in 15 results which were further explored (Lenau et al. 2011).

Table 4-5: Exemplary process of search term formulation (reported by Lenau et al. 2011)

Technical task	**Reduce the consequences of car collision**
Initial search terms	*Collision, shock absorption, absorb energy and control deceleration*
Brainstorming → Additional search terms	*impact AND shell, protection layer, protection mechanism, energy absorption and impact AND protection'*
After initial search in library and on asknature.org	Partly unsuitable results due to double meanings → Search term combinations, *e.g. protection AND impact*; Use of synonyms, e.g. for *collision*: *clash, conflict, smash..*

This example already shows some strategies for formulating search terms. In the following subsections, we present three methods which can be used for systematic search term formulation. To start with, we present methods for translation between the technical and the biological domain (4.4.1, fact sheet p. 195). Then, we explain the variation of search terms (4.4.2, fact sheet p. 197) and the

BIOscrabble method (4.4.3 Fact sheet p. 199) which proposes the use of additional non-functional search terms.

Figure 4-13 compares the suitability of the three methods for different bio-inspired design scenarios and approaches. It shows that the methods are suitable for all presented scenarios and both biology push and technology pull approach.

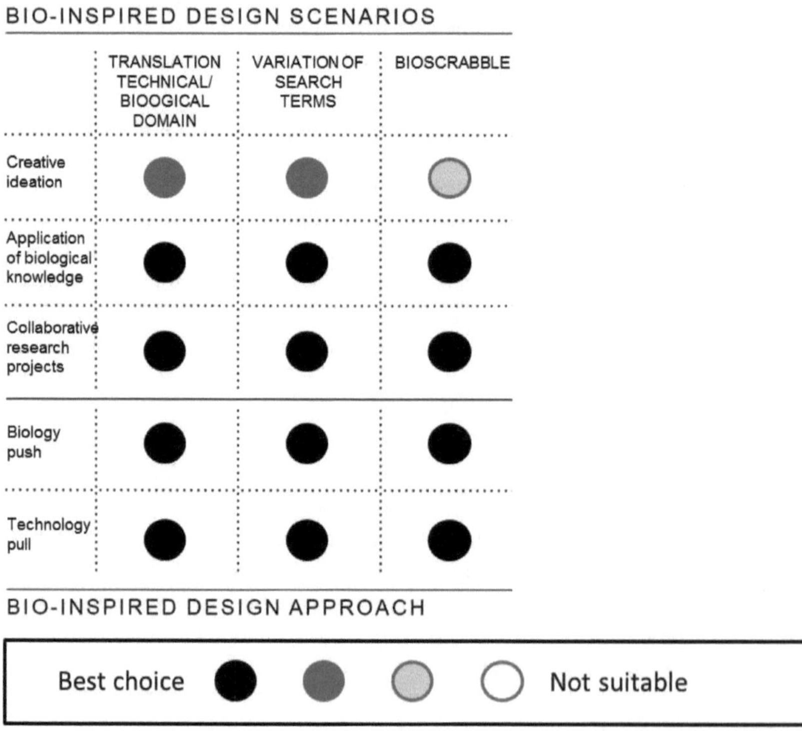

Figure 4-13: Suitability of search methods for different bio-inspired design scenarios and approaches (how to search?)

BEST PRACTICE

Bio-inspired Bone Punch – Searching for Partial Functions

Fraunhofer Institute for Manufacturing Engineering and Automation IPA

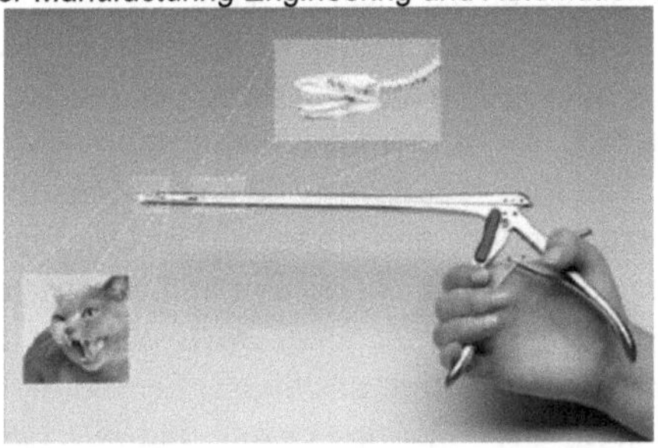

photo: © Fotodesign Rolf Schwarz, Ludwigsburg

A bone punch is a surgical instrument for the surgical extraction of bone and cartilage. It must fit into narrow openings, cut pieces of bone or cartilage, and transport them out of the wound. The precise cutting of cartilage and bone material, as well as the transportation through the long narrow opening, are technically challenging. To develop the bio-inspired bone punch, the designers at the Fraunhofer Institute (IPA) formulated medical search terms such as *resected tissue* and technical search terms such as *punch*. The search terms were translated to the biological domain by a technical biologist and varied. This resulted in the terms *parts*, *particles* and *meat cuts* for *resected tissue* and *separate*, *cut,* and *bite* for *punch*. Using these terms, a systematic search could be conducted. Moreover, the biological translations themselves hinted at possible biological inspirations: the term *meat cut* could be directly related to carnivores. Consequently, the designers focused on carnivores and their carnassial teeth. As a result, the punching mechanism was inspired by cats: their carnassial teeth slide over one another as scissors. The bone punch's punching mechanism was modified so that its blade slides positively over an anvil. The best inspiration for the transportation mechanism was found during an excursion of the design team to a natural historic museum. There, the designers spotted the skeleton of a snake. Snakes have a similar problem to a bone punch – they have to transport their prey through their long bodies. The design team chose the anaconda as biological model and closely analyzed the anaconda's swallowing mechanism. The anaconda swallows its prey by alternating upward and downward movements of upper and lower jaw. As well as this, its teeth point in the downward direction so that the prey cannot move upward during the swallowing process. The designers transferred this mechanism and integrated a rail with teeth pointing outward. The rail moves inward and outward to transport the cut bone or cartilage material.

Based on an interview with the group manager at Fraunhofer IPA, Oliver Schwarz (2016), and Schwarz, Miklosovic (2013), Fraunhofer-Institut für Produktionstechnik und Automatisierung IPA (2015), Machata (2016)

4.4.1 Translation Technical/Biological Domain

Fact sheet p. 195

Several taxonomies have been developed to translate technical search terms to biological search terms. They can also be used the other way around: to translate biological search terms to the technical domain.

We present three taxonomies that have been developed in the last ten years. The complete taxonomies are included in the appendix (section 9.3, p. 253).

On the technical side, all of them use the terminology of the *Functional Basis* (Hirtz et al. 2002). The *Functional Basis* contains terms for abstract functions and flows. Using abstract functions and flows, i.e. material, energy and signal flows, designers can abstract their technical system and construct a flow-oriented model (see section 3.1.1, p. 59). The *Functional Basis* contains terms for functions and flows on three levels of abstraction or *classes*. For example, the function *connect* (primary class) contains the functions *couple* and *mix* (secondary class). *Couple* contains the functions *join* and *link* (tertiary class). Moreover, a number of synonyms are defined for each function. Similarly, the flow *material* (primary class) can be a *human, gas, liquid, solid, plasma, or mixture* flow (secondary class). The solid flow can be an *object*, a *particulate,* or a *composite* flow (tertiary class). Again, synonyms are listed for the flows (Hirtz et al. 2002).

On the biological side, the three taxonomies differ, as they have been developed using different approaches. Therefore, the provided "translations" also differ. Table 4-6 shows an example: the technical terms connect (or couple/join/link) are translated to a number of different biological terms, only the List of Biologically-meaningful Keywords and the Engineering-to-Biology Thesaurus provide partially corresponding translations (overlap, bind, activate). This is probably due to the fact that the Engineering-to-Biology Thesaurus uses terms from the List of Biologically-meaningful Keywords.

Table 4-6: Overview taxonomies for translation technical-biological domain

	List of Biologically-meaningful Keywords (Cheong et al. 2011)	Engineering-to-Biology Thesaurus (Nagel et al. 2010)	Correlation matrix NIST-BT (Baldussu 2014)
Source of technical terms	functional basis (Hirtz et al. 2002)		
Source of biological terms	biological textbook *Life* (Purves et al. 2001)	functions: list of biologically meaningful keywords, Idea-Inspire database, flows: biological textbook *Life* (Purves et al. 2001)	biomimicry taxonomy (The Biomimicry 3.8 Institute 2016)
Result	matching between biological terms and functions of the Functional basis	matching between biological terms and functions and flows of the Functional basis	matching between the Functional basis and the Biomimicry Taxonomy
Example: Biological translation of the technical functions *connect/couple/join/link*	*extend, project, hold, stretch, overlap, activate, bind*	*recombination, mate, build, phosphorylate, bond, synthesis, latch, lock, extend, link, overlap, bind, adhere, fuse, clamp, activate, project*	*attach permanently/temporarily, physically assemble structure*

To start with, the **list of biologically meaningful keywords** has been developed by Cheong et al. (2011) based on an analysis of the biology textbook *Life* (Purves et al. 2001). Verbs from the functional basis were taken and regrouped according to the lexical database WordNet® (version3.0) (Princeton University 2010). For example, the function *prevent* (tertiary class) is a higher level function in the WordNet® database and several other terms of the functional basis can be assigned to it (*inhibit, protect, shield*). Using the verb groups, the biology textbook *Life* was searched for frequent nouns in the vicinity of the technical terms and collocated *bridge* verbs (Cheong et al. 2011).

Taking the technical term *protect* as an example, Cheong et al. (2011) explain the following sentence they took from the textbook:

"Within the shell and surrounding the embryo are membranes that protect the embryo from desiccation..."

According to Cheong et al. (2011), *embryo* is one of the frequent nouns and *surround* is a collocated bridge verb.

The collocated bridge verbs were then searched for in a biology dictionary. If there was a definition for such a verb, it was considered *biologically meaningful*. In cases where the verb was only used in other definitions, it was considered *biologically connotative* (this is the case for surround). Cheong et al. (2011) determined if a biologically meaningful or connotative verb was too general or too specific by considering its overall frequency in the biology dictionary. If this was not the case, the verb was included in the list of biologically meaningful keywords for technical functions of the functional basis. Moreover, for each biological term, Cheong et al. (2011) calculated the ratio of collocations with the technical term and the total number of matches in the biological textbook. Based on these numbers, they recommend using the keywords that have a low collocation ratio and a high number of matches only when searching in combination with the technical term. Otherwise, these frequent but unspecific keywords can yield high numbers of irrelevant search results (Cheong et al. 2011).

To address technical functions, the **Engineering-to-Biology-Thesaurus** (Nagel et al. 2010; Nagel 2014) uses the biological terms extracted by the authors of the list of biologically meaningful keywords (Cheong, H., Shu, L. H. et al. 2008) and terms used in the Idea-Inspire database. The Idea-Inspire database presents biological systems modeled according to the SAPPhIRE approach (see section 3.4.3, page 76 and fact sheet VIII, page 183) (Srinivasan and Chakrabarti 2009). Additionally, the Engineering-to-Biology-Thesaurus provides a translation of the technical flows proposed by the functional basis. The biological translation was extracted by searching the biological textbook *Life* (Purves et al. 2001). Biological nouns in the vicinity of the technical functions were only analyzed if they appeared in at least two cases: if we take the sentence cited above as example, the terms *embryo* and *membrane* are possible candidates. These nouns were cross-checked in a dictionary and assigned to the technical flows. The sample terms, *embryo* and *membrane,* are regarded as material flows (primary class). *Embryo* is placed under *object* (secondary class) and *solid* (tertiary class). Membrane is placed under *mixture* (secondary class) and *solid-solid mixture* (tertiary class) (Nagel et al. 2010; Nagel 2014).

The **Correlation matrix NIST-BT** (Baldussu 2014, p.86-87, 226-230) takes a different approach: biological terms are not extracted from a biological source, but the terms for groups, subgroups, and functions of the Biomimicry taxonomy (The

Biomimicry 3.8 Institute 2016) are used. The functions and flows of the functional basis are assigned to the subgroups and functions of the Biomimicry taxonomy. The approach (starting point: biological domain) is therefore inverse to the List of Biologically Meaningful Keywords and to the Engineering-to-Biology-Thesaurus (starting point: technical domain). The result of this assignment is not complete, as the functional basis does not contain correspondents for every subgroup and function of the biomimicry taxonomy. For example, no functional basis term was found for the subgroups *coordinate* and *cooperate and compete* (Baldussu 2014, p.86-87, 226-230).

To conclude, all translation taxonomies can be used both for biology push and technology pull approaches. Using the taxonomies, the search in both domains can be improved by adapting the search terms to the specific domain. This facilitates search particularly for non-experts. However, as Table 4-6 shows, the List of Biologically Meaningful Keywords and the Engineering-to-Biology Thesaurus in particular provide numerous translations. For a search, keywords still have to be selected to make the search results manageable. The translation taxonomy can be used for creative ideation, application of biological knowledge, and collaborative research projects. For creative ideation, the additional time effort has to be considered.

Using the translation taxonomies can be well combined with searching for research, publications, technical patents and online databases of biological principles. In particular, the Correlation matrix NIST-BT is tailored to be used for searching in the asknature.org database (Deldin and Schuknecht 2014).

4.4.2 Variation of Search Terms

Fact sheet p. 197

Instead of using taxonomies for translating technical search terms to the biological domain or vice versa, variations of search terms can be built.

Several types of variations have been suggested (e.g. by Shu and Cheong 2014). Variations can be built using Wordnet®, a database of the English language that groups words according to their semantic and lexical relations (Fellbaum 1999; Princeton University 2010). An alternative is Verbnet which includes verbs only and additionally uses syntactic information to classify them (Kipper et al. 2000; University of Pennsylvania 2005). We propose the following variations of search terms (examples are built using Wordnet® by Princeton University 2010)

- Synonyms: To increase the space of search results, synonyms, i.e. terms with the same meaning as the original term, can be used. For example, regarding the climbing glove task, a designer could formulate the search term *adhere* and build synonyms such as *stick, cling, cleave* and *cohere*

- Troponyms: These are a specification of the original search term, e.g. to *mold* means to fit tightly or to follow the surface of an object. It is a specific manner to *adhere* to an object.
- Hypernyms: These are generalizations of the original search term, e.g. to *touch*, *adjoin*, *meet* and *contact* are hypernyms of *adhere*
- Sister terms: Hypernyms have several troponyms. Therefore, words are linked to sister terms with a slightly different meaning via their hypernyms. Sister terms of adhere are for example *attach*, *hug* and *rub*. The verb *attach* is in this case also an *entailment*, i.e. if the glove *adheres* to a surface, it *is attached* to the surface.
- Antonyms: An antonym is a word with the opposite meaning of the original term. For adhere, Wordnet® does not list an antonym. The antonym of *attach* is *detach*.
- Different parts of speech: Without changing the meaning of a word, it can be varied by building different parts of speech. For the verb *adhere*, different parts of speech are the noun *adhesion* and the adjective *adherent*. In a search, different parts of speech can often be included by using abbreviations, for example the search term *adhe** which includes *adhere*, *adhesion* and *adherent* (Helms 2016).

The variation of search terms can be used both for biology push approaches to search for technical application areas and for technology pull-approaches to search for biological solution ideas. It is useful for creative ideation if the additional time-effort is acceptable for the application of biological knowledge and collaborative research projects. It can be combined with any search resources – research publications, technical patents, catalogs, or databases. On the downside, Helms (2016) concluded from results from an empirical study that the variation of search terms was relatively ineffective and time-consuming.

4.4.3 BIOscrabble

Fact sheet p. 199

The majority of the search methods in bio-inspired design are based on searching for functions. This is due to the manner of abstraction in technical product development – designers often formulate their technical problem using functions. Alternatively, they may abstract technical systems to formulate functions on highly abstract levels. However, as we observed in chapter 3, functions do not play such a prominent role in biology. Helms (2016) therefore bases her work on the assumption that in addition to functions, searching with other terms must be useful. She proposes to search using terms for functions, properties, environments, and other concepts:

- Function: A function is a verb that describes the purpose of the technical system (input-output relationship) (Kaiser et al. 2014). Regarding the climbing glove example, one function of the glove is to adhere to surfaces.

- Property: Attribute of the technical system (Kaiser et al. 2014), usually described by an adjective or adverb, e.g. the gloves should only adhere temporarily to a surface.
- Environment: Description of the surroundings and their influence on the technical system (Kaiser et al. 2014), e.g. the slippery rocks play a role in the glove adhesion. Environment can also refer to other technical components, for example technical components of the system that should not be modified due to a new design.
- Other Search Terms: Additional search terms that do not fall into the three categories *function*, *property* and *environment* can be useful as well. An example is a description of a function carrier, for example the glove is designed to be worn by hands, so search terms such as *hands* or related limbs such as *fingers*, *feet,* and *toes* can be useful.

In the next step, variations of the search terms can be built as described in the last sub-sections. However, as mentioned there, Helms (2016) concluded from a study that search term variations were not very effective.

To conduct the search for biological publications, the publication database PubMed (National Center for Biotechnology Information 2018) is recommended. The database includes more than 27 million citations for publications from biology and medicine. As we know from other internet searches, a search can be conducted with combinations of several search terms. Helms (2016) recommends to:

- Specify searches, but not make them too specific so as not to limit the search space. Especially if quotation marks are used, all results that do not include the terms in the same order are excluded.
- Combine search terms via Boolean operators (*AND, OR, NOT*)
- Combine search terms describing properties and environment with other types of search terms (e.g. property AND environment, property AND function etc.).

Helms (2016) also gives recommendations for dealing with the search results:

- Sort results: The search adjustments can be used to sort the results. On PubMed, the standard sorting is according to topicality (*most recent*). Another option is to sort the results according to relevance (*best match*).
- Reduce results to a manageable number: As a search often results in a high number of matches, the results have to be filtered. Helms (2016) does not recommend automatic filtering, nor does she recommend to discard search results due to a title. Even if a designer does not understand the title of a publication, it is worthwhile to regard it in detail. An option to reduce search results by randomly choosing publications for a closer analysis (e.g. every fifth result)
- Analyze publications: At least the abstract of a publication should be read. To make the decision more objective, KO criteria can be defined previously. The evaluation of the publications should be documented in a list.

A considerable advantage of *BIOscrabble* is that it pushes designers to reflect on their technical task from several angles and not to focus too much on functional descriptions. Moreover the method includes recommendations on how to combine search terms and to deal with the search results that are useful for other search methods as well. As research publications are used, *BIOscrabble* allows access to a large search space and the latest publications. On the downside, the analysis of the search results takes time, particularly because the publications can be difficult to understand for non-experts.

The *BIOscrabble* method has been developed for technology pull approaches and is tailored to use the online database PubMed. Nevertheless, as other search engines and online databases have similar functionalities, *BIOscrabble* can be used as well. It can also be adapted to search for technical patents for biology push approaches. In this case, search terms are based on functions, properties and the environment of a biological system.

As the use of research publications (and patents) requires considerable time-effort, *BIOscrabble* is less suitable for creative ideation. It can be used just as well for the application of biological knowledge and collaborative research projects.

5 Analysis and Comparison

Why is analysis and comparison of biological and technical system important enough to dedicate a whole chapter to it? In many bio-inspired design processes such as the Technology Pull Procedure (fact sheet p. 169) and the Biology Push Procedure (fact sheet p. 171) *Analysis and Comparison* is not an explicit stage.

Analysis and comparison seem to be too self-evident to explicitly demand them as activities. But this might be too shortsighted – cognitive psychologists have found that experts spend more time analyzing a problem than novices. Comparing different systems, they find the less obvious but more relevant similarities. For example, a study conducted by Chi et al. (1981) presented physical problems to novices and experts. The experts categorized the problems according to the underlying physical principles necessary for a solution. The novices categorized the problems with regards to the objects they contained such as springs and inclined planes (Chi et al. 1981; Goldstein 2011, p.347).

It is therefore worth spending time analyzing and comparing biological inspiration and technical problems before actually transferring bio-inspired analogies. It is impossible to present the whole range of analysis methods for both disciplines in this chapter. To give an idea of common approaches, we introduce several standard methods in technical product development (5.1) and biology (5.2). For both disciplines, we differentiate between physical and virtual analysis methods. In section 5.3, we compare these approaches and deduce possible synergies in bio-inspired design projects. Then, we introduce five methods for analysis and comparison in bio-inspired design (5.4).

5.1 Analysis of Technical Systems

If we situate the analysis of technical systems in product development processes such as the *Munich Procedural Model* (see Figure 2-4, p. 25), we find them mainly in a stage of *properties assessment* to prepare the evaluation of developed technical solutions. Additionally, existing technical systems can be analyzed for *goal planning* and *analysis* (Lindemann 2009, p.45–49).

To assess a technical system's properties, different types of variables are considered: input variables and output variables (see Figure 5-1). Input variables are defined by the designers. Examples for input variables include shape, material, geometric dimensions, and so on. Output variables fall into the categories of functional variables and relational variables. Functional variables include all performance variables related to the function of the technical system, such as the speed of a car, or the mechanic stability of a single component. Relational variables originate from the interaction of the technical system to other technical and non-technical systems. Examples include manufacturability, costs, and environmental impact (Göker 1996; Lindemann 2009, p.160; DIN 2330: 2013-07, 2013; DIN 2330: 2013-07, 2013).

H. Hashemi Farzaneh and U. Lindemann, *A Practical Guide to Bio-inspired Design*, https://doi.org/10.1007/978-3-662-57684-7_5

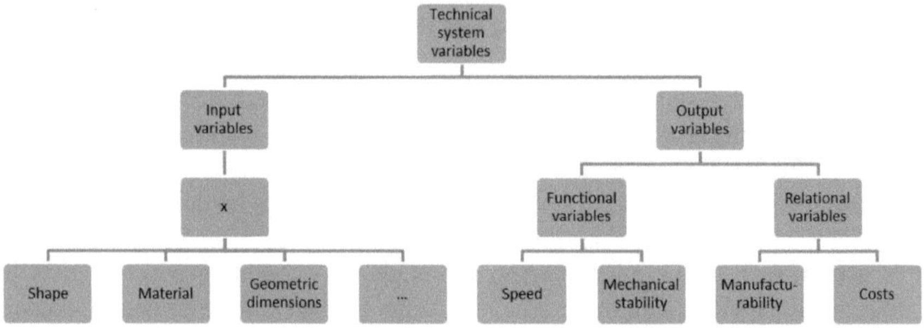

Figure 5-1: Technical system variables (based on Göker 1996; Lindemann 2009, p.160)

To summarize, the input variables of technical systems are, unlike the properties of biological systems, predefined by the designers, and can be used as adjusting levers to influence functional and relational variables. In most cases, only these output variables have to be assessed for the analysis of technical systems.

The analysis process can then be simplified as shown in Figure 5-2: in addition to the predefined input variables, disturbance variables play a role. Disturbance variables describe all known and unknown influences. For example, the temperature differences between winter and summer and night and day cause the elongation and shortening of technical and building components. Designers have to consider this and provide expansion gaps in bridges, roads, and so forth. The analysis process records all relevant output variables. According to the measurements, input variables can be adjusted to achieve the required output variables (Lindemann 2009, p.162).

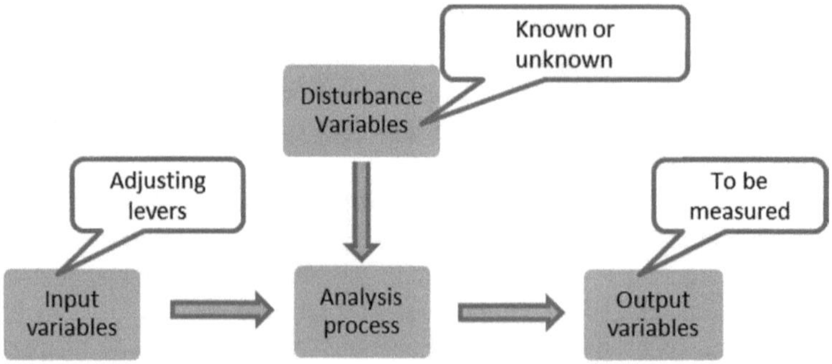

Figure 5-2: Technical analysis process (Lindemann 2009, p.162)

Which concrete methods are used for the analysis of technical systems?

Throughout the last decades, virtual engineering has advanced enormously. In the middle of the last century, designers developed products at drawing boards or tables. This completely changed to using CAD (Computer-aided Design) technologies at the end of the century. Instead of drawing models of technical systems and their components, designers represent them virtually in CAD models. In the same period, numerical simulation methods were developed. Using these CAE (Computer-aided Engineering) methods to analyze CAD models, designers can predict technical system properties (Bauer and Dangelmeier 2016). CAD models are virtual prototypes, i.e. they approximate an interesting aspect of a technical system. Virtual prototypes are particularly flexible – their parameters can be changed by a few "clicks" in computer software (Ulrich and Eppinger 2016, p.293–303).

However, physical prototypes are still needed to test technical system properties in reality. The detection of unexpected phenomena is not possible without physical prototypes (Ulrich and Eppinger 2016, p.293–303). They range from very simple prototypes in the early development phases to exact reproductions of the final product.

In recent years, both virtual and physical prototyping and testing have evolved due to technological advancements which enable entirely new analysis approaches: virtual reality enables a direct interaction between the human and the virtual model of a technical system (Bauer and Dangelmeier 2016). For physical analysis, additive manufacturing technologies allow the so-called 3D-printing of components with very few geometric restrictions (Beyer and Kochan 2013; Ulrich and Eppinger 2016, p.304).

In the following subsections, we will introduce both virtual and physical prototyping.

5.1.1 Virtual Analysis

Besides CAE methods, several other approaches exist to virtually assess the output variables shown in Figure 5-1:

- CAPE: Computer-aided Process Engineering, such as simulations of assembling processes
- CAM: Computer-aided Manufacturing, such as simulation of a milling process
- CAQ: Computer-aided Quality Management, such as simulation of manufacturing tolerances

These approaches focus on the relational variables depicted in Figure 5-1 concerning manufacturing, environment, and so on. In contrast to these approaches, CAE methods focus on the functional variables. Moreover, CAO (Computer-aided Optimization) links CAE and CAD, numerical simulation and design. CAO applies numerical simulation to optimize CAD models of technical system components (Rieg et al. 2016).

In this subsection, we focus on CAE methods, as the assessment of functional variables is crucial for bio-inspired design. Typical functional variables that designers evaluate are (Schäfer 2006, p.1–2):

- Stresses and deformations of components (solids)
- Flow velocities and pressure
- Temperature distribution in solid components or fluids
- Rates of heat or mass transfer
- Pressure or energy losses

These properties serve to assess crucial functions of technical products. For example, determining the temperature distribution in ovens is necessary to assess the quality of baking pastry. Calculating the Eigen frequency of bridges is important for predicting and preventing collapse due to winds. Knowledge of the deformation of automobiles in crash situations is necessary to improve passenger safety (Schäfer 2006, p.1–2):

As technical systems and their components are usually too complex to calculate these properties analytically, CAE methods are based on numerical calculations. For numerical calculations, complex components are broken down to small elements with known properties. Based on these small elements, numerical calculations provide approximated solutions for the overall technical component or system. To calculate solutions with sufficient precision, computers are needed for the so-called numerical simulation. Both computation capacities and the development of numerical algorithms since the second half of the 20th century were drivers for the wide-spread application of numerical simulation (Schäfer 2006, p.4–5; Rieg et al. 2016).

Figure 5-3 shows a general procedure for numerical simulation proposed by Schäfer (2006, p.7):

In the first step, the technical system must be modeled using an adequate mathematical model. This requires a step of abstraction and simplification of the actual technical system. For example, a common model is the beam model. For model beams, several assumptions are made. One of these assumptions is that model beams can be cut into slides which are flat and vertical to the beams axis even in a deformed state. Real beams do not fulfil this prerequisite exactly. However, using the model beam to approximate real beams results in good approximations of reality if the deformation of the beam is relatively small (Weichert 2005; Schäfer 2006, p.33). In addition to the technical system or components, the relevant forces and contacts to other objects have to be modeled as well.

The results of this abstraction step are differential equations describing the model and boundary conditions (e.g. small deformations for the beam model).

In the next step, the continuous model must be transformed into a discrete model, or, in other words, it has to be divided into a finite number of elements. This process is called discretization. In practical computer applications, a grid can be generated on the components to generate a finite number of elements. The finer the grid, the more accurate the numerical simulation is. However, the computing time also increases (Schäfer 2006, p.57–75).

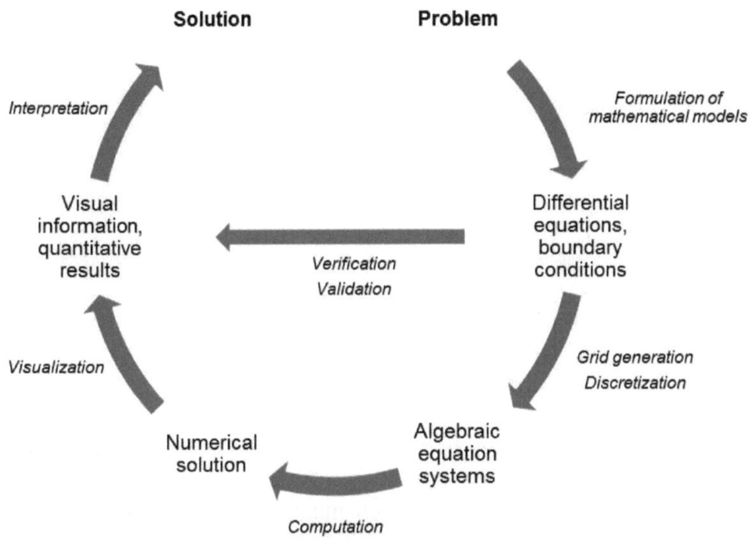

Figure 5-3: Virtual analysis: Procedure for numerical simulation (based on Schäfer 2006, p.7)

Then, the resulting algebraic equation systems can be numerically computed. Modern software for numerical simulation gives numerous options to visualize the results. An example is shown in Figure 5-4: stresses are highlighted with different colors – from blue (low) to red (high). Deformations are exaggerated to make them visible. The results have to be analyzed carefully to ensure that the initial model was suitable (validation) and that the equations were solved correctly (verification). After an initial simulation, the model of the technical system and its components or the discretization often has to be reconsidered. If the result of the simulation is validated and verified, it can be interpreted to make assumptions on the functional performance of the technical system (Schäfer 2006, p.9).

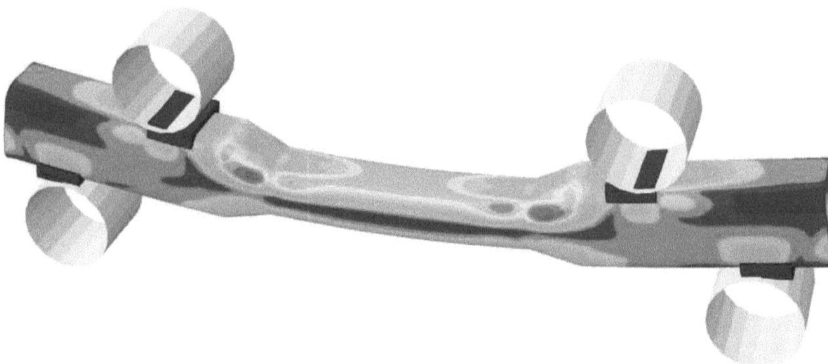

Figure 5-4: Result of a numerical simulation of a braided carbon fiber reinforced plastic beam (the cylinders represent the force transmission points), image: J. Achleitner/TUM (Achleitner 2014)

The technological advances of the recent decades have enabled an enhanced representation of numerical simulation results via *Virtual Reality* tools. *Virtual Reality* describes the interaction between humans and a realistic digital product model –designers or users immerse themselves into an artificial digital environment in which they can directly interact with the product model. The high degree of realism is achieved via 3-D models, sound effects, and in some cases also haptic feedback. Virtual Reality requires expensive technical equipment (e.g. projectors, head-mounted displays, eye-tracking…) and computation capacities. The computation time for simultaneous FEM simulations is at present too high to be used in practice. Nevertheless, simulation results can be visualized in *Virtual Reality* environments to make them tangible (Lashin 2013; Bauer and Dangelmeier 2016).

In addition to the display of numerical simulation results, *Virtual Reality* can be used for numerous other analysis purposes (Bauer and Dangelmeier 2016):

- Reviewing the geometric and aesthetic design
- Testing the feasibility of the technical system with regards to manufacturing, assembly, and maintenance
- Exploring the usability and ergonomic aspects of the technical system

To summarize, Ulrich and Eppinger (2003, p.296) classify analysis methods along an axis of comprehensiveness. They regard CAE methods as highly focused in comparison to physical analysis methods, which can also be comprehensive. *Virtual Reality* has the potential to shift this limitation and allow a more comprehensive virtual analysis of technical systems.

5.1.2 Physical Analysis

Physical prototypes are very common means for analysis in technical product development. With the emergence of CAE methods, physical prototypes have been replaced by virtual prototypes to some extent, especially in the early stages of prototyping. In recent years, this has changed due to the emergence of added manufacturing, rapid prototyping, and 3D-printing technologies. All these terms describe technologies that construct components in a layering technique: software cuts a virtual CAD model into flat slices or layers of equal thickness. These slices are then applied by a rapid prototyping machine or 3D printer. The material is applied as liquid, powder, foil, or wire. Today, numerous materials can be used, such as polymers, ceramics, metals etc. A similar technique is even used to *print* biological cell structures. However, this technique is still in the early development stages: So far, it has been used to print and transplant different types of tissue, such as skin, bone and heart tissue (Murphy and Atala 2014).

For technical applications, the polymers/ceramics/metal prototype can be solidified using laser, UV light, binder, or glue. In the final step, the finished prototype has to be post-processed. Depending on the concrete additive manufacturing technique, supporting structures must be cut out of the prototype, porous prototypes must be infiltrated with stabilizing material, and so on. An additional option is not to build the actual prototype using additive manufacturing, but instead to build molds for other material which is more suitable for the final prototype (Beyer and Kochan 2013; Schmutzler et al. 2016; Ulrich and Eppinger 2016, p.304)

To conclude, the advantages of additive manufacturing techniques for building physical prototypes are:

- Good integration into virtual product development processes due to the potentially rapid drafting of physical prototypes based on a CAD model
- Ability to build complex shapes with cavities
- Depending on the material used, they are useful to assess functional variables and provide a haptic and visual model of the technical system.

Figure 5-5 depicts an example of a component manufactured with additive manufacturing technologies.

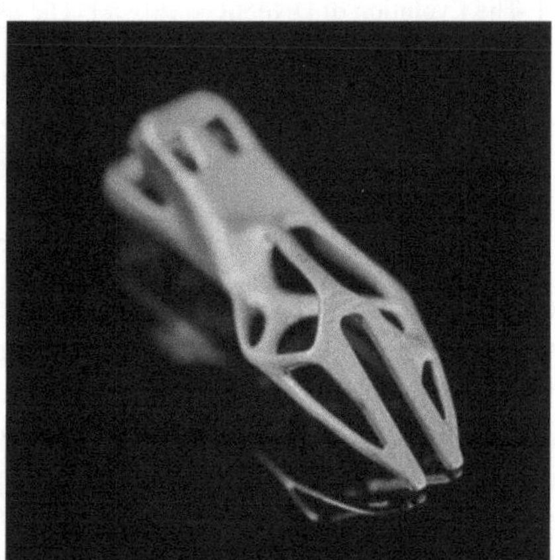

Figure 5-5: Bio-inspired 3D-printed component (fixture in the tonneau cover for the soft-top) for the BMW i8 Roadster (photo: BMW)

It is a mass-produced 3D-printed component for the BMW i8 Roadster. The component was optimized analogous to natural shapes (BMW Group 2018). For such complex shapes, comparatively expensive additive manufacturing technologies can replace machining processes even in mass production.

5.2 Analysis of Biological Systems

The science of biology incorporates the analysis of living systems on an enormous scale, from the level of molecules to that of the entire earth.

Campbell et al. (2018, p.52–53) list ten levels of biological organization: molecules, organelles, cells, tissue, organs, organisms, populations, communities, ecosystems, and the biosphere. Regarding two introductory biological textbooks (Purves et al. 2004; Campbell et al. 2018), Table 5-1 shows that more sections are dedicated to the smaller levels – molecules, organelles and cells – than to the larger levels of biological organization. Moreover, the research methods illustrated in both textbooks mostly focus on these levels (Purves et al. 2004, back *page: Directory of research methods,* Campbell et al. 2018, p.25: *Research methods figures*).

Table 5-1: Levels of biological organization in biological textbooks

Levels of biological organization	Sections of *Life* (Purves et al. 2004)	Sections of *Biology* (Campbell et al. 2018)
Molecules, organelles, cells	The Cell	The Role of Chemistry in Biology
	Information and Heredity	Cell Biology
	Development	The Genetic Basis of Life
	Evolutionary Processes	Evolution
	The Evolution of Diversity	The Diversity of Life
Tissue, organs, organism	The Biology of Flowering Plants	Plants: Structure and Function
	The Biology of Animals	Animals: Structure and Function
Population, communities, ecosystems, biosphere	Ecology and Biogeography	The Ecology of Life

A possible explanation is that, with increasing level of organization, biological systems become more complex. On each level of biological organization, new properties emerge resulting from the interaction of system elements on the lower levels of biological organization. To understand biological systems, biologists therefore start on the smaller levels. (Campbell et al. 2018, p.73)

Another aspect is the similarity of biological systems on the levels of molecules, organelles, and cells. For example, four types of macromolecules can be identified in all biological systems in approximately the same proportions: proteins, nucleic acids, carbohydrates, and lipids (Purves et al. 2004, p.37).

Findings on biochemical and microbiological level can therefore explain biological phenomena on large scale. An example is genetics: hereditary information that determines biological systems is stored in the macromolecule deoxyribonucleic acid (DNA). By analyzing chromosomes, the threadlike structures composed of DNA, many biological phenomena on the higher levels of biological organization can be explained. For example, different species can be differentiated by their numbers and sizes of chromosomes (Alberts et al. 2015, p.173–193).

In the following subsections, we will introduce some basic biological approaches to analyzing biological systems. Both virtual analyses ("dry experiments") and physical analyses ("wet experiments") allow the examination of the different levels of biological organization.

5.2.1 Virtual analysis

Computional models can represent a wide range of biological phenomena, such as shape and structure, interaction and development, and evolution of biological systems and their element. There are therefore few standard computational models (that would, for example, be comparable to technical FEM models). The general procedure for setting up a computational model of a biological system is shown in section 3.2.2 (p. 64). Here, we explain three examples of virtual analysis and prediction to give an idea of the diversity:

An example for the virtual analysis of **shape and structure** is the simulation of virus' shells (capsids). Viruses are among the smallest biological systems in nature. Their capsids consist of protein structures. A well-known case is the simulation of the capsid of the HIV virus. Even with an electron microscope, the HIV capsid structures cannot be clearly visualized. Proteins can be bred to give an idea of the protein structures, but they are still not equivalent to real capsids. Zhao et al. (2013) used for example data from electron microscopy and artificial protein structures. Simulating 64 million atoms with a supercomputer, they were able to visualize the variations of the capsid of the HIV virus. This finding reveals the "weak spots" on the surface of the HIV virus, i.e. spots at which drugs can dock and incapacitate the virus (Bajaj 2007; Zhao et al. 2013).

Interaction between biological systems and their elements can for example be simulated using graphs. Graphs consist of elements (nodes) connected by relations (edges) which can be either directed or undirected. Functional models in technical product development, for example, are a type of graph representation. In biology, graphs can be used to simulate all kind of networks, from very small to very large scales. A few examples (Fontoura Costa 2007):

- On a very small scale, proteins can be represented as elements, compatibility between proteins as relations. Docking between proteins can then be simulated.
- On a larger scale, neuronal networks are a very typical application for graphs: single neurons and synaptic relations between neurons can be represented. Moreover, whole cortical areas and the connections between areas can be modeled.

- On an even larger scale, interaction between organisms, such as predator and prey in an ecosystem, can be simulated

An example for a simulation of development and evolution is the simulation of the development of parasites' resistance to insecticides. Onstad and Meinke (2010) simulated the development of resistance of the western corn rootworm in different scenarios for planting transgenic corn. For the simulation, they used a computational model of the worm's population genetics and a model landscape. The results indicate how to manage the plantation of transgenic corn to prevent a fast development of resistances of the worm (Onstad and Meinke 2010).

These few examples illustrate the variety and potential of virtual analysis in biology. Nevertheless, virtual analysis requires computational models which highly simplify and abstract complex biological systems. This increases the risk of mistakes and incorrect predictions or analysis results.

5.2.2 Physical Analysis

If we step down from the high levels of biological organization to the detailed levels, then different methods have to be applied to, firstly, break down a biological system into its elements and, secondly, observe these elements. We can for example identify a single horse in a group of horses and observe its behavior. But if we want to collect its genetic information and analyze it, we need technical equipment. Table 5-2 shows examples for different preparation and observation methods on several levels of scale (Campbell et al. 2018, p.52–53).

To start with, we can observe, count, and mark species large enough to be visible to the naked eye on the high levels of biological organization, such as ecosystems, communities, populations, and organisms, Smaller biological organisms such as bacteria, for example, must be observed using microscopy. An example is the adhesion strategies of tree and torrent frogs. Experiments can be conducted on this level as well. This is exemplified by experiments with tree and torrent frogs adhering to a glass plate with varying inclination. Besides the maximum angle at which a frog can adhere to the plate, these experiments have also revealed the frogs' strategies for adhesion: if the angle becomes very steep, frogs spread their legs to optimize the contact angle.

Table 5-2: Examples for preparation and observation on different levels of biological organization (Levels of organization according to Campbell et al. 2018, p.52–53)

Level of biological organization	Identification, preparation	Observation
Biosphere *The entire earth*	-	(Via satellites from space)
Ecosystem *Biological species, environment*	Counting Marking …	Observation with naked eye
Community *Biological species in an ecosystem*		
Population *Group of individuals of a species*		
Organism *Individual of a species*		
Organ *Body part with specific functions*		
Tissue *Group of cells fulfilling functions*	Fixation, Embedding/Freezing, Slicing	Light or electron microscopy
Cell *Structural and functional unit*		
Organelle *Functional cell component*		
Molecule *Chemical structure of several atoms*	Cultivation, Multiplication and decomposition of cell cultures (e.g. by centrifugation)	Chromatography, Electrophoresis, Spectroscopy…

Stepping down from the levels of organs to tissues and cells, structures become too small to be visible to the naked eye. Figure 5-6 depicts the use of light and electron microscopy to observe biological systems on different scales. The naked eye is able to see objects as small as a few hundred micrometers. Consequently, to analyze, objects as small as human skin tissue, light microscopy is required. Light microscopes are an arrangement of optical lenses that focus an image of an illuminated specimen into the user's eye. The properties of light waves determine the limit of resolution of light microscopy: ~200 nm for conventional resolution and up to ~20 nm for super-resolution. This enables the analysis of animal and plant cells and bacteria. To observe tissue under a microscope, it usually has to be sliced into thin sections. To protect the cells in the tissue, biologists use fixatives. As tissue is often too soft to be cut directly it is additionally embedded in wax, resin, or another supporting material. Another option is to freeze the tissue. The now solid tissue can be cut into slices with a thickness of typically 0.5 to 10 μm. Using fluorescent chemicals, light microscopy can detect much smaller elements than the limit of resolution allows: specific molecules can be marked fluorescently and made visible despite their small size (Alberts et al. 2015, p.529–554).

To observe organelles and molecules, electron microscopy enables a resolution of up to 0.05 nm. Electron microscopes use electrons instead of light to produce an image of the specimen. The use of electron microscopy requires to fixate, dehydrate, and embed biological tissue to be observed. The embedded specimens must be cut into slices with a thickness of 25 to 100 nm. To prevent damage, biological tissue is often rapidly frozen and then embedded. Moreover, to make molecules visible, the tissue has to be prepared or "stained" with salts of heavy metals (Alberts et al. 2015, p.554–562).

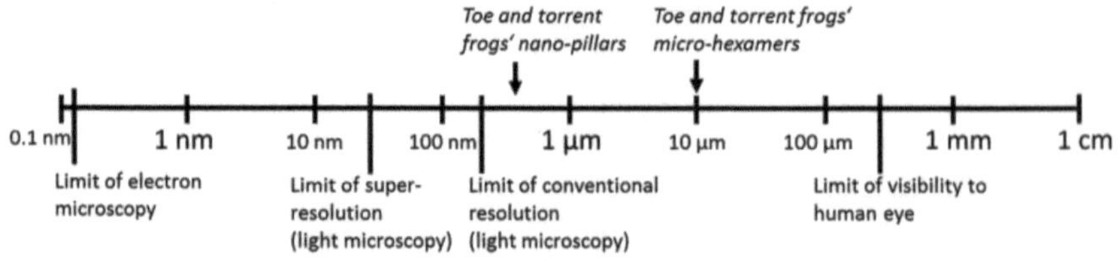

Figure 5-6: Microscopy for analyzing biological systems on small scales (Alberts et al. 2015, p.531)

In addition to observing organelles and molecules from samples of tissue, chemical analyses enable the biochemical processes within a cell to be understood. Biochemical methods often require large numbers of specific cell types. Therefore, cells must be isolated from tissue samples. Then, they must be cultivated, multiplied, and decomposed into their components (Alberts et al. 2015, p.439–508).

To isolate cells, tissue is prepared with a chemical solution which dissolves the bonds between the cells. Cells can be sorted according to different cell types, e.g. by using fluorescent markers and a so-called cell sorter. Cells can then be cultivated on culture dish where they divide and multiply. Cell cultures can also be used to test the effect of medicine, hormones, and other such things. As many cells stop dividing after a certain number of iterations, established cell lines are used for many experiments with cell culture. Common cell lines have been extracted from model organisms such as the fruit fly, mice, or the tobacco plant (Richter 2003, p.5–23; Alberts et al. 2015, p.439–508).

To decompose cells into their components, they are mechanically broken up, e.g. by vibration, pressure, grinding, or blending. The result is a homogenate containing the organelles of the cell. Then, these cell components must be separated. Centrifugation at increasing speeds separates cell components into molecules according to their size and molecular mass. To separate components of similar size according to their density, a solution with a density gradient is used (velocity or equilibrium sedimentation). The cell components can be chemically analyzed using techniques such as chromatography, electrophoresis, and spectroscopy. This allows to identify for example proteins or DNA and to draw conclusions to the chemical processes within cells. This understanding supports the explanation of biological processes on the cellular and chemical level, such as information processing, reproduction, and so on. (Richter 2003; Alberts et al. 2015, p.439–508)

5.3 Comparing Technical and Biological Analysis Methods

The previous subsections have shown similarities and differences of the analytical approaches in technical product development and biology. Figure 5-7 is a tentative approach to visualizing similarities, differences, and possible synergies.

To start with, the different starting points of technical product development and biology play a crucial role in analysis. In particular, the standard approaches to physical analysis differ considerably:

In technical product development, the starting point is a technical task or problem. For this task or problem, designers develop concepts which are initially vague. To analyze them physically, prototypes have to be made, at the beginning often with very simple tools such as paper or building blocks (Brown 2008). The more the development process advances, the more sophisticated the prototypes become, and they increasingly resemble the final product. This necessitates more sophisticated tools, for example traditional workshop tools such as turning laces or drilling machines, or additive manufacturing technologies. The physical prototypes are often tested for their resistance to external influences, e.g. external forces or temperature, and for the fulfilment of their main functions. In addition to these physically measurable factors, physical prototypes are used to evaluate "soft" factors such as ergonomic aspects or aesthetic design.

In biology, the starting point is a highly complex biological system. Physical analysis therefore implies the observation or examination of the biological system itself. To be able to examine the biological system on different levels of biological organization, preparation is often required, such as marking of individuals or cultivating cells. To observe or examine the biological system, light and electron microscopy is required for the tissue to molecular level. Moreover, different chemical analyses are common, for example spectroscopy methods (Alberts et al. 2015, p.439–564).

As Figure 5-7 depicts, rather "technical" methods for making and testing can be beneficial for analyzing biological systems. Technical systems can be manipulated and tested much easier than living biological systems. By this means, bio-inspired design can entail synergy effects for biological research. A good example is research into tree frog adhesion: Drotlef et al. (2013) prototyped variations of the tree frog's microstructure. Using these prototypes, they tested normal and shear forces under different conditions of wetness. From the results, they were able to draw conclusions as to the function of the microstructure of tree frogs' toe pads: the microstructure increases shear forces and therefore its function is to facilitate climbing. Another example is the underwater adhesion of the terrestrial leaf beetle: the terrestrial leaf beetle is able to walk on surfaces underwater. To understand its adhesion mechanism, Hosoda and Gorb (2012) reproduced the beetle's setae (fine hair) with silicone polymers. Using the silicone prototype, they were able to inject air bubbles between the silicone setae and could test their effect on adhesion. They found that the air bubbles increased adhesion to hydrophobic surfaces.

Regarding virtual analysis methods, CAE approaches, in particular numerical simulations, are widely used in technical product development. Comparable numerical simulations are also applied in biological research (e.g. for biomechanics as proposed by Maas et al. 2012). For biological systems, setting up a virtual model requires more abstraction, as they are more complex than technical systems. This also increases the risk of a less valid model and consequently less useful analysis results. In both disciplines, the use of virtual reality approaches is still restricted to particular cases. This is mainly due to the high technological effort and its tendency to be time-consuming. In technical product development, virtual reality tools are used for example in automotive design, aerospace design, or plant engineering (Lashin 2013). In biological research, virtual reality allows researchers to display and interact with large amounts of data. One example is a three dimensional model of the brain - researchers can discuss and analyze it in a virtual reality environment (Winkels 2017).

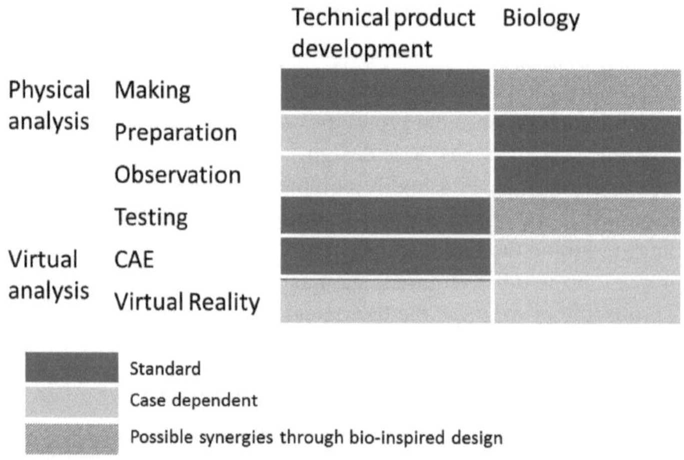

Figure 5-7: Comparison of technical and biological analysis methods

BEST PRACTICE

Vault-structured Materials – Looking beyond the Obvious

Dr. Mirtsch Wölbstrukturierung GmbH

We have all heard of the miraculous properties of hexagonal structures in nature – the stability of honeycombs, the delicateness and perfection of snowflakes, the deformation and fluid-absorption of the microstructure of the torrent frog's toes – just to name a few examples. However, if we simply try to mimic the hexagonal structure using conventional manufacturing processes, the result might not be as expected: for example, metal sheets with an embossed hexagonal structure lose strength, as the embossing process induces stress onto the material, causing material thinning. We must look beyond the obvious to reproduce the properties of nature's hexagonal structure. Natural structures are not the result of forced deformation, but instead they emerge through self-organization. A comparable process which was initially considered to be undesirable was observed for metal cylinders. In the 1930s, small dents were observed on the surface of submarines that had been exposed to very high water pressure. Later, in the 1970s, a researcher accidentally set metal thin-walled cylinders under moderate external pressure. The cylinders that contained inner rigid stiffening rings deformed: staggered square patterns emerged on their surface. The researcher's curiosity was awoken and F. Mirtsch started studying the self-organization of natural structures. In 1992, he and two partners founded the company Dr. Mirtsch GmbH. During this venture, he discovered the emergence of hexagonal structures on cylinders by using flexible instead of rigid stiffness rings. This is the "Basic Vault-Structuring-Process". It is a type of structuring by self-organisation and entails batch-wise production, because only single cylinders can be structured. To enable mass production, the Dr. Mirtsch Wölbstrukturierung GmbH developed a modified, continuously-working manufacturing process for hexagonally or 3d-faceted vault-structuring in metal sheets, papers, cardboards, or other materials, such as CFK. These vault-structured materials are characterized by an increased all-round bending-stiffness and decreased buckling behavior, including the preservation of the surface quality of the initial materials. The vault-structured materials yield enhanced flow-characteristics when overflown with fluids, resulting in an enhanced heat- and mass-transfer. Consequently, the material thickness can be decreased to reduce weight. Today, vault-structured materials have been used for numerous applications. For example, due to their high stability in crash tests, they are used for automotive components. Moreover, sound isabsorbed, which makes vault-structured material very suitable for floor panels. Vault-structured sheets are also used for a washing drum which treats delicate clothes with particular care (Miele & Cie. KG). Another interesting property is the reflection of light. Vault-structured metal has therefore been used for an innovative lighting system: the Hexal LED trunking system (Osram/ SITECO GmbH).

Based on an interview with founder F. Mirtsch (September 2017 and Mirtsch, Mirtsch (2012)

5.4 Methods for Analysis and Comparison in Bio-inspired Design

In the previous subsections, we have introduced typical analysis approaches in technical product development and biology. Even though there are common methods and possible synergy effects through bio-inspired design (see Figure 5-7), the analysis methods are mainly domain specific. Consequently, in a collaborative research project, biologists are probably responsible for the analysis of the biological inspiration, and engineers for the testing of the bio-inspired technical system. Therefore, there are few suggestions for **analysis** approaches particular to bio-inspired design projects.

A guideline for *Material Analysis* summarizes methods on different scales for bio-inspired design projects (VDI 6223, 2013). *Material Analysis* generates new knowledge and requires considerable effort; therefore, it is only applicable in collaborative research projects. Moreover, we describe a method which draws from the consultation of experts from biology and technical product development – the Expert-Novice-Layperson Dialog (Jordan 2008, p.118–122). In direct communication, knowledge from the experts can contribute both to the analysis and the comparison of biological and technical systems. This method is highly recommendable to initiate collaborative research projects, but it can also be used for other bio-inspired design scenarios. In particular, the composition of a group of experts and novices can be interesting for creative ideation.

Figure 5-8 shows three further methods which particularly address the **comparison** of biological and technical systems:

Analysis of Scaling Effects evaluates the comparability of fluid flows, heat transfer and forces across scales. *Analysis of Scaling Effects* should be conducted whenever the scale of a biological solution is changed to be applied in a bio-inspired technical system. Only in case of creative ideation on a superficial level, the required information for *Analysis of Scaling Effects* is possibly not available.

In contrast to *Analysis of Scaling Effects* with predefined equations, the following two methods can be used to define and reflect on comparison criteria:

The *Decision Model Similarity* proposes comparing functions, constraints and quality criteria of biological and technical systems (Zerbst 1987, p.28–30). To adequately compare these factors, detailed understanding of biological and technical system is needed. The method is therefore suitable for collaborative research projects and for the application of biological knowledge if enough details are available. The *Four box method* is used to define criteria for comparison (Yen et al. 2014). It requires fewer details than the other methods and is therefore useful in all bio-inspired design scenarios, especially for the application of biological knowledge.

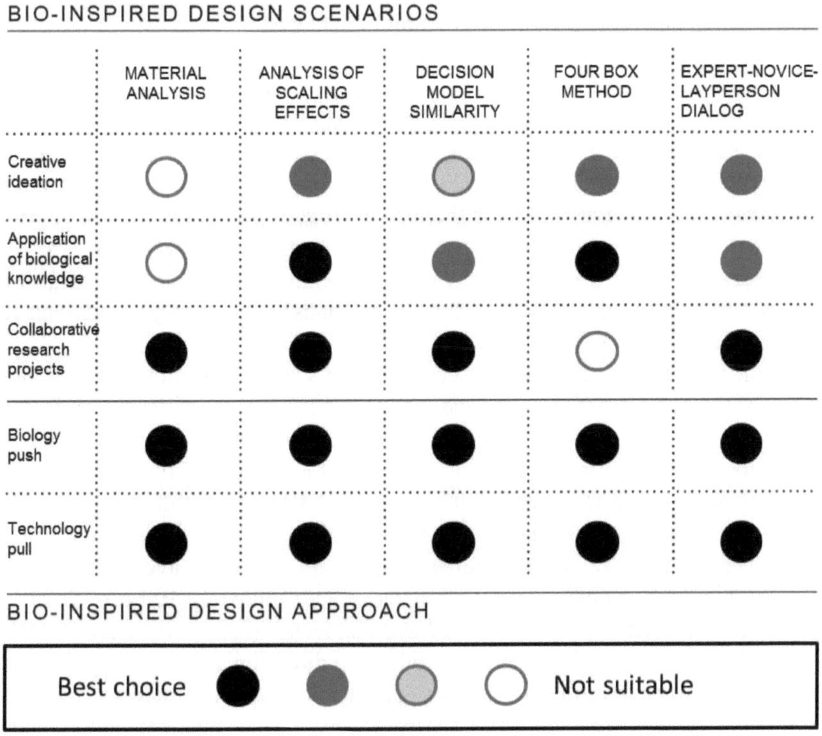

Figure 5-8: Suitability of analysis and comparison methods for different bio-inspired design scenarios and approaches

5.4.1 Material Analysis

Fact sheet p. 201

As explained in section 5.2, the hierarchical structure of a biological system plays a crucial role in its behavior and functions. On each level of biological organization, material and structure fulfil specific functions (see Table 5-2, p. 125). To understand how a biological system fulfils overall functions, researchers have to carefully analyze each hierarchical level and the interplay between them.

The torrent frogs' toe structure is a good example for the importance of the different hierarchical levels (see also Figure 1-4, p. 14). It successfully adheres to wet, slippery, almost vertical rocks. We know that, on the torrent frog's toes, both pillars on the nano-level and hexamers on the micro-level contribute to adhesion. On every scale, there are several underlying principles that enable adhesion under different conditions.

Material Analysis (2013) gives an overview on the basic techniques to examine different hierarchical levels of biological systems:

On the largest scale, the biological system can be examined on the **integral level**, i.e. as a whole. On this scale, a biological system such as a torrent frog can be observed in its natural habitat. Moreover, mechanical or behavioral tests can be conducted. For example, researchers have tested the adhesion capabilities of torrent frogs by setting them on a glass plate. They slowly turned the plate and observed the angle at which the frogs could not adhere anymore and subsequently fell off the plate (Endlein et al. 2013b). The overall geometry and large surface patterns can be analyzed. As for the torrent frog, at this point we can consider its physical build: it enables the frog to press itself effectively against almost vertical surfaces. Moreover, the frog can spread its legs and toes. Tree frogs grip small twigs using their toes – several adhesive toe pads support its gripping capabilities (Endlein et al. 2017).

From this observation, we descend by one scale to the **macroscopic level** of tissues. For the torrent frog, we can analyze the structure of its toe pads. On this scale we can still conduct mechanical tests and analyze surfaces with the naked eye. However, more useful is a common light microscope. Conventional light microscopes enable to see objects with a minimum size of 0.2 μm.

As such, they can also be used on the next smaller scale, the **microscopic level**. Here, we can analyze cell structures. Typical animal cells have a diameter of 10-20 μm, typical plant cells have a diameter of approximately 100 μm. To compare, the micro-hexamers of torrent frogs have diameters of approximately 10 to 15 μm (Endlein and Barnes 2014). In order to build a three-dimensional model of the biological specimen, *tomography* is used. This technique is well-known from medical CT scans. Images are taken from different views of the specimen and combined to a three-dimensional model (Alberts et al. 2015, p.557–558).

On the **submicroscopic level**, below the limit of resolution, electron microscopy is applied. The principle of electron microscopy is similar to that of light microscopy. However, electrons are used instead of light. A transmission electron microscope (TEM) consists of an electron gun and a combination of lenses to focus the electrons on the specimen. TEM is used for two-dimensional images of the specimen, whereas *Scanning Electron Microscopy* (SEM) can provide three-dimensional images. There are further specific techniques, e.g. Multiphoton imaging is a light microscopic technique using infrared laser light to analyze living tissue (Alberts et al. 2015, p.529–564).

On the **biochemical level**, electron microscopy can still be used to visualize molecules. To analyze the chemical composition of biological substances, spectroscopic methods are used (e.g. FT-IR, Raman, UV, NMR). Spectroscopy is based on the principle that different molecules absorb different wave-lengths of radiation. The peaks of the spectrum of a substance indicate different components. Even on the biochemical level, molecules can be manipulated and forces can be measured using *Atomic Force Microscopy* (AFM). A sharply pointed tip is attached to a springy cantilever arm mounted on a detector. Using the tip, the AFM can measure the mechanical properties of single protein molecules (Richter 2003, p.99–124; Alberts et al. 2015, p.529–564).

5.4.2 Analysis of Scaling Effects

Fact sheet p. 203

Both in biology and in the technical domain, the importance of dimensions is well known. If an engineer wants to build a smaller prototype of an airplane to test the behavior of a real airplane in a wind channel, they have to consider non-linear fluid forces. Similarly, non-linear physical relations influence biological systems. In comparison to larger mammals, a mouse needs much more energy to keep its body temperature constant, as its body surface is much larger relative to its mass.

In bio-inspired design, dimensions have to be taken into account very frequently. The torrent frog is a good example – if we want to transfer the structure of its toe pads to technical applications, we might face challenges when searching for suitable technical materials and manufacturing techniques. The hexamers and pillars measure no more than a few micro- or nanometers. Depending on the concrete application and available material, these small scales can be difficult to reproduce. Therefore, we possibly have to re-dimension them and reproduce their structure on a larger scale. How does this influence their behavior? Does the complex interplay of different physical effects still work on a larger scale?

The fact sheet *Analysis of Scaling Effects* gives an overview on three relevant aspects to consider when changing the scales of a biological solution to apply it in the technical domain.

To start with, **fluid flows** have to be considered. The viscosity of fluids such as air and water do not increase linearly with the size of immersed objects. As a consequence, the forces within flow fields with objects of different sizes are not comparable. The Reynolds-Number was developed to ensure comparability of geometrically similar flow fields with different sizes: it describes the ratio between inertia and friction. Flow fields with the same Reynolds-Number behave similarly. In the case of the model airplane, its smaller mass results in a lower Reynolds-Number than that of a real airplane. A strategy to allow experiments with the model airplane is to reduce the viscosity of the fluid to decrease friction, i.e. by cooling the air in the wind tunnel. The Reynolds-Number is also used to differentiate between laminar and turbulent flows. If the Reynolds-Number exceeds 2300, the fluid flow is considered turbulent. Laminar and turbulent flows behave entirely differently: on a macroscopic level, laminar flows consist of parallel layers, whereas turbulent flows are characterized by random transient vortices. A major difficulty of using the Reynolds-Number is to choose the *characteristic length* of the immersed objects. If the fluid flows around the immersed object (e.g. the wing of an airplane or the body of a fish.), the length of the object in the direction of flow is considered. If the object poses a barrier to the flow, the width of the object is taken a as *characteristic length*. This differentiation requires experience and it is recommendable for non-experts to compare the calculations to similar applications and to consult experienced designers (Zerbst 1987, p.169; Zierep and Bühler 2015, p.127–129).

Another aspect that we must consider when changing the scales of a biological solution is **heat transfer**. Heat transfer occurs through conduction and radiation.

Regarding conduction, we must differentiate between (1) conduction between solids (or stationary fluids), (2) free convection between moving fluids, and (3) forced convection between fluids moving due to an external stimulus and the resulting pressure difference, for example (Böckh and Wetzel 2015, p.3–4).

The rate of **conduction** between solids (1) depends on the surface area divided by the material thickness. If we therefore scale up a surface area and keep the material thickness constant, the rate of conduction will increase. The rate of convection due to fluid flows (2 & 3) depends on multiple variables such as dimensions and Reynolds number (for details see for example Baehr and Stephan 2011).

Unlike conduction, **radiation** does not require contact between materials. It describes contactless heat transfer due to electromagnetic waves from one object to another. An example is the heat transfer from sun to earth. Radiation rates depend on the surface area. If we scale up a surface area, heat transfer through radiation increases.

Moreover, the influence of mechanical **forces** differs considerably on different scales. On large scales, gravity plays a more important role than on small scales, whereas on small scales, friction forces have more influence. Moreover, on small scales, intermolecular forces such as *Van der Waals* forces are important.

In the torrent frog example, friction, *Van der Waals* forces, and the capillary effect contribute to the adhesion capabilities of the frog. If we scale up the frog's toe pad structure, the influence of friction and *Van der Waals* forces decreases. The capillary effect is inversely proportional to the width of the spaces between the hexamers on the micro-level and the pillars on the nano-level. If we scale up the hexamer- or the pillar-structure, the capillary effect decreases unless we keep the spaces between the hexamers or pillars small.

This example shows that the transfer of biological solutions to other dimensions has to be evaluated carefully, and the calculations shown here are useful to estimate scaling effects in a simplified manner. The main challenge is to identify all relevant aspects of the biological solution that depend on the dimensions. Nevertheless, in the second step, consultation of experts, more accurate calculations and simulations, and the testing of technical prototypes can be necessary.

Analysis of Scaling Effects has to be conducted whenever we change the dimensions of a solution; both for biology push and technology pull approaches, for collaborative research projects, the application of biological knowledge, and less often for creative ideation. If we have to scale the biological solution up or down, the *Decision Model for Similarity* is a suitable method for transferring analogies to the technical domain.

5.4.3 Decision Model Similarity

Fact sheet p. 205

The Decision Model Similarity supports the comparison of biological and technical systems. If they differ in crucial aspects, Zerbst (1987) proposes abandoning the regarded biological system and searching for more suitable ones. The Decision

Model Similarity includes three steps for evaluating the similarity between a technical and a biological system (Rechenberg 1978; Zerbst 1987, p.28–30):

Firstly, the similarity of technical and biological functions is considered. Zerbst (1987) regards function from a technical perspective, i.e. it is defined as the purpose to be achieved by a structure, shape, or principle of a technical or biological system. Taking the torrent frog example, its microstructure has (amongst other functions) the purpose to absorb and emit fluid. This function can be considered as similar to the technical product idea of a plaster which applies medicine: the plaster should absorb fluids from the wound and emit medicine.

Secondly, technical and biological constraints are examined. Constraints describe all invariable quantities that have an impact on the technical or biological systems. They can be a consequence of the external environment in which the technical system must function, such as external loads, flow forces, wind, or sun. However, the constraints can also stem from technical manufacturability and feasibility, such as available material and manufacturing techniques. In this step, scaling effects play a crucial role and designers have to analyze the biological system and the technical system carefully (more details on *Analysis of Scaling Effects* have been explained in section 0, p. 132, fact sheet p. 203).

With regards to the torrent frog and the medicine plaster, designers have to evaluate whether the absorption capacities of micro-hexamers work under the constraints of the medicine plaster: the fluid is an important external factor: the torrent frog absorbs and emits water and secretion. Can the hexamer structure absorb and emit both blood and medicine – fluids with entirely different properties? The material is a relevant internal factor: can the micro-hexamers be built using technical materials which are suitable for wound contact?

Thirdly, if a detailed analysis has provided a positive answer to these questions, the quality criteria are regarded: what properties of the technical solution influence its performance? What properties of the biological system influence its chances for survival? Unlike constraints, properties are considered as variables; designers can influence the technical systems quality criteria. Regarding the medicine plaster, the emission and absorption rates play a role: the material with micro-hexamer structure is in contact with the wound and must absorb a certain amount of blood and other bodily fluids. Moreover, it must emit the required rate of medicine. Variables influencing absorption and emission capacities are the position of the plaster relative to the wound and pressure on the hexamer structure. The frog's capacities to absorb and emit fluid depend on the wetness of the environment and on the pressure due to the frog's weight. It must also be evaluated whether the touch of the user's hand on the medicine plaster imposes enough pressure on the hexamer structure for the required medicine emission.

Zerbst (1987) suggests designing a bio-inspired solution only if functions, constraints, and quality criteria are can be considered similar. Otherwise, designers should look for different biological inspirations. This strict limitation can cause a too hasty abandonment of biological systems. Before abandoning a biological inspiration, designers should think about alternative options such as changing the abstraction level of an analogy.

Nevertheless, the method has several advantages, in particular for collaborative research projects and the application of biological knowledge: the *Decision Model Similarity* draws attention to constraints and quality criteria. Differences can make bio-inspired solutions fail, especially if the technical and biological systems have different dimensions. Moreover, the structured, detailed analysis of similarities and differences is very useful. For creative ideation, the knowledge of the biological or technical system is in most cases probably not sufficient to conduct the method. The Decision Model Similarity can be used both for biology push and technology pull projects.

5.4.4 Four Box Method

Fact Sheet p. 207

Observations in student classes inspired the development of the *Four Box Method*: Yen et al. (2014) observed that student teams working on bio-inspired design tasks regularly failed to justify their decision for a certain biological inspiration. Even though the students had a preference for a specific inspiration, they had difficulties giving concrete reasons for this preference.

The *Four Box Method* was therefore introduced to provide structure to the comparison between a biological and technical system. It proposes four aspects for consideration: environment, function, specification and performance. Designers must list properties of the considered biological and technical system in a table. In the next step, they compare these properties using the terms *same*, *similar,* and *different* (Yen et al. 2014).

The fact sheet (p. 207) shows a sample comparison of the technical system *climbing glove* and the biological system *torrent frog*.
It is illustrated in Figure 5-9:

The *environment* aspect refers to the operating environment of the technical system and the habitat of the biological system. For example, designers must develop climbing gloves for use in temperate regions (10-25°C), and climbing wet rocks. The torrent frog species lives around waterfalls in temperate mountain regions (15 °C). In this case, the operating environment of the climbing glove and the habitat of the torrent frog can be considered *similar*.

The *functions* of the climbing glove include the support of the climbing person; they should therefore hold a quarter of the climbing person's weight. The torrent frog's toe pad and skin surface have a *similar* function; they have to hold the entire weight of the frog.

Specification aspects describe properties of the technical and biological systems which are instead qualitative and cannot be quantified. An example is the force to adhere and detach the climbing gloves to a rock: the climber should only have to exercise low forces to adhere and to detach his hands. For torrent frogs, we observe the same specification – torrent frogs seem to exercise low forces to adhere to rocks and detach from them to move forward.

Performance aspects describe quantifiable properties of technical and biological systems. If we translate the abstract function "to *hold* (a portion of) the climbing person's weight" to concrete numbers, the gloves must resist a certain tensile stress. Considering a person with 80 kg and a contact area of 10x10 cm per glove, we can calculate a tensile stress of 1 N/cm^2. In comparison, the torrent frog only has to resist 0.2 N/cm^2. The performance is therefore different. This comparatively low performance of the torrent frog makes the transfer of its solution principles to a glove design questionable as the design might not meet the requirements.

An advantage of *Four Box Method* is to highlight these critical issues by a side-by-side comparison of biological and technical systems. Moreover, it supports structuring biological and technical knowledge. The main challenge of the method is acquiring reliable information on the technical and biological systems.

The Four Box Method can be used equally for biology push and technology pull-approaches. As its application is simple and requires no considerable time-effort, it can be used in all bio-inspired design scenarios, but is most suitable for the application of biological knowledge.

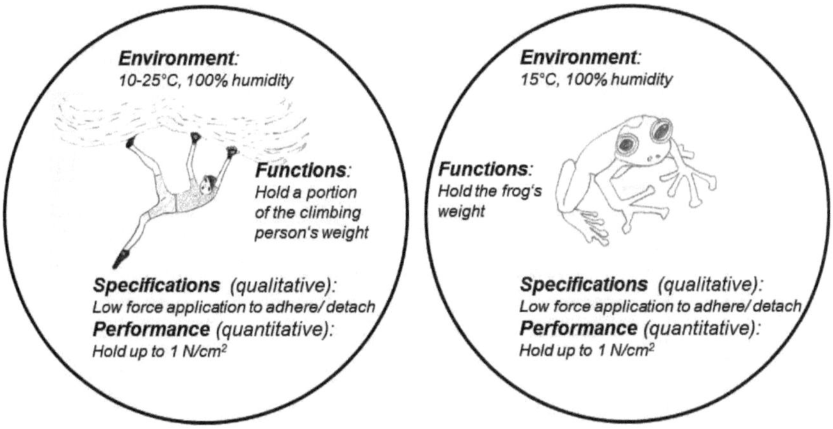

Figure 5-9: Four Box Method example

5.4.5 Expert-Novice-Layperson Dialog

Fact sheet p.209

In contrast to the previously presented methods, the expert-novice-layperson dialog supports interpersonal knowledge transfer. In a dialog between a biologist and an engineer, technical and biological systems can be effectively analyzed and compared. If the biologist is an expert on the biological system and the engineer is an expert on the technical system, existing knowledge is available on the spot.

This is a considerable advantage in comparison to approaches that rely on searching for documented information. Nevertheless, knowledge transfer between experts from different disciplines can also be challenging: Jordan (2008, p.118–122) argues that one main challenge is the large knowledge gap between experts from different disciplines. Even though typical experts possess profound knowledge in their discipline, they lack knowledge in other disciplines. They can therefore be considered *laypersons*. This can be illustrated as shown in Figure 5-10: there is only a small overlap between the biologists' and the engineers' knowledge that can serve as a basis to start a dialog. An additional aspect is the lack of common terminology or language. This is a challenge for other activities of bio-inspired design as well, for example when searching: technical search terms differ from biological terms used to describe biological systems.

To overcome the knowledge and language gap, Jordan (2008, p.118–122) proposes a novice as a mediator. According to Jordan (2008), the inclusion of such a novice changes the dialog entirely. As shown in Figure 5-10, the novice has to possess a medium level of knowledge in both disciplines. He or she therefore shares more knowledge with both experts. Moreover, the novice is familiar with some biological and technical terminology. Nevertheless, the level of knowledge and terminology is lower than the expert level. Consequently, the novice can ask suitable questions to bridge the knowledge gap between the two experts. Jordan (2008, p.118–122) emphasizes that the novice should not be involved in the bio-inspired design project. In such a case, they would possess too much knowledge in one of the disciplines.

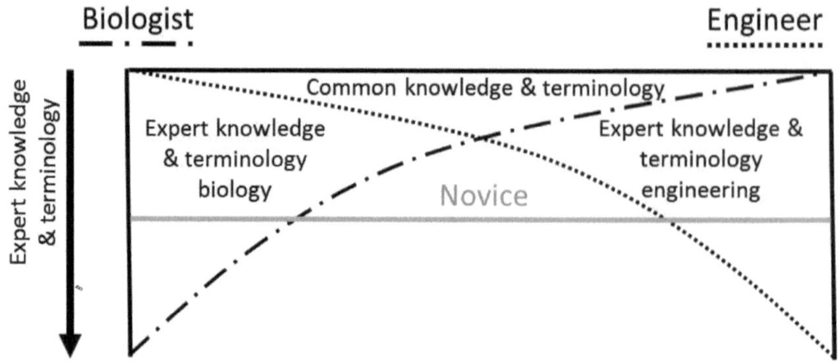

Figure 5-10: Small knowledge overlap between experts in biology and technical product development (adapted from Jordan 2008, p.118–122)

Despite the described advantages of the method, the major challenge is finding a suitable novice and preparing them for the role in the dialog.
Jordan (2008, p.118–122) proposes students for the role of the novice. However, ideally, the novice possesses knowledge in both disciplines. Biology or engineering students are therefore less appropriate for the role unless they have enrolled in classes from the other discipline or classes on bio-inspired design. Another possi-

bility could be biologists or engineers with experience in other bio-inspired design projects. Fulfilling the role of the mediator can also be a challenge of the expert-novice-layperson dialog. Especially if the novice is a student, the experts possibly have reservations towards them. It can be difficult for the novice to actively mediate between the two experts.

The expert-novice-layperson dialog can be applied both for biology push and technology pull projects. It is most recommendable to start collaborative research projects, but it can also be used for the application of biological knowledge. It is also promising to conduct creative ideation in the biologist expert – novice – technical expert constellation.

BEST PRACTICE

From biopolymer Fibers to Filter Systems – Collaborative Analysis with Research and Industry Partners

University of Bayreuth, Technical University of Munich, Bosch und Siemens Hausgeräte GmbH, Sandler AG

Spider silk has fascinated researchers for a considerable amount of time; it is highly tear-resistant, durable, elastic, and lightweight. Spider-silk consists of biopolymers – namely proteins. Silk proteins from different spider species have been analyzed and the encoding DNA sequences have been identified. By inserting these DNA sequences into bacteria, proteins can be produced recombinantly. The produced biopolymers can be, but are not necessarily an exact copy of, natural spider silk proteins. These artificially produced proteins must be spun to produce fibers.

The BioFis project aimed to integrate the bio-inspired spider silk fibers into vacuum cleaner filter systems. As the silk fibers are electrostatically charged, an increase in the adhesion of charged dust particles was expected. The project was conducted with two university laboratories (biomaterials and product development) and two companies (vacuum cleaners and filter systems). At the time of the project, research on bio-inspired polymer fibers was ongoing and no large quantities could be produced. This turned out to be challenging for the project, as in order to develop the filter system, research was needed on several levels; i.e. on the production of the spider silk fibers, on the integration into a fleece, and on the linkage to a carrier material. This research could often not be conducted in parallel. Consequently, project partners had to wait for each other's results. Nevertheless, the early involvement of the different project partners allowed for a target-oriented approach on all levels. For example, in comparison to basic research without defined application, the filter application constraints reduced the variables that had to be considered for processing spider silk into fibers and non-woven fleeces. The project resulted in a filter with a spider silk layer that collects fine dust at levels comparable to that of a commercially available system, but with much higher air flow, meaning a much less energy consumption.

Based on an interview with researcher H. Bargel (October 2017), and Helten et al. (2012), BIOKON (2014)

6 Transfer of Analogies

The transfer of analogies and the generation of bio-inspired technical solutions is the goal of bio-inspired design. At a first glance, bio-inspired analogies are often fascinating – gecko-inspired robots which climb walls, swim suits mimicking a shark skin structure, artificial spider-silk… The technical solutions stemming from a bio-inspired analogy seem to be ultimately creative and certainly exotic in comparison to non-bio-inspired technical products.

But, in fact, analogical transfer is neither a genius skill nor a complex method, it is part of our everyday thinking process: cognitive psychologists have found that we use analogies all the time. For example, Dunbar (1999) observed weekly laboratory meetings in biological research groups. He found that researchers used analogies to formulate hypotheses, design or fix experiments and explain results. In each one-hour laboratory meeting, between three and fifteen analogies were formulated (Dunbar 1999; Dunbar and Blanchette 2001). Using the same method in an engineering company, Christensen and Schunn (2007) counted on average 11 analogies per hour in meetings of a design team.

Even though analogical transfer seems to be a natural cognitive approach to tackle problems, researchers have reported a number of difficulties related to the transfer of adequate analogies (Goldstein 2011, p.340–346). In this chapter, we explain a general cognitive perspective (6.1) and then switch to the analysis and support of analogical transfer for designing technical products (6.2). Based on this theory, we introduce six methods which support the analogical transfer in bio-inspired design (6.3)

6.1 General Perspective on Analogical Transfer

In a general sense, why and how do we use analogies?

A common strategy used to face unfamiliar situations is to think about comparable, familiar situations. Often, we can solve problems arising from the unfamiliar by transferring knowledge from the familiar. For example, we are facing the problem of having to repair an air mattress. We can easily come up with a solution even if we have never repaired an air mattress before: we immediately think of the process of repairing a flat tire, a problem any user of bikes has probably faced multiple times.

This knowledge transfer is analogical transfer – it is depicted in Figure 6-1: The bike represents the familiar situation or domain. We know a solution to a familiar problem: the flat tire. The air mattress is a new, unfamiliar situation or domain. In this situation we do not have experience with solutions. The two situations can be regarded as analogs: they are characterized by "horizontal" (i.e. "similarity") relationships and "vertical" (i.e. "causal") relationships between two aspects of one of the analogs (Hesse 1970, p.59).

H. Hashemi Farzaneh and U. Lindemann, *A Practical Guide to Bio-inspired Design*, https://doi.org/10.1007/978-3-662-57684-7_6

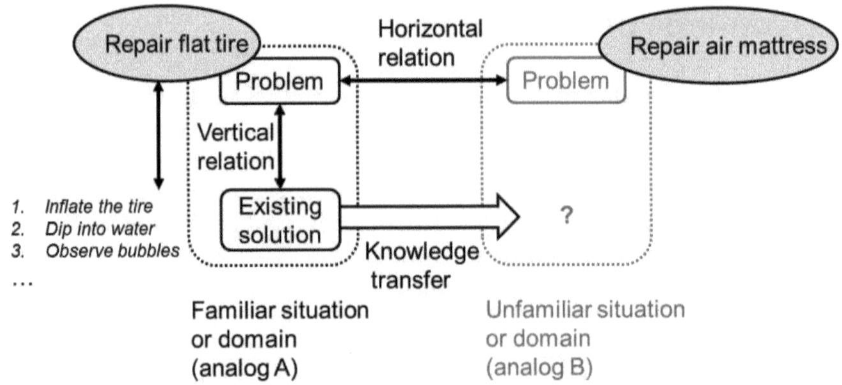

Figure 6-1: Analogical transfer – from tire to air mattress

The analogical transfer for problem solving entails three steps (Gick and Holyoak 1983; Goldstein 2011, p.340–346):

- *Noticing*: We must first notice the analogical relation between the two analogs: the horizontal relation. Regarding our example, both tire and air mattress are soft, inflatable objects.
- *Mapping*: In the second step, we must identify the corresponding elements of the two analogs, e.g. the similarity of the hull of a tire or an air mattress. Therefore, the reason why they lose air is probably the same: there is a hole in their hull has a hole.
- *Applying*: In the last step, we apply the solution of the familiar analog to the unfamiliar analog. In many cases, we have to adapt the solution to make it suitable. Regarding our example, we know that if we immerse the inflated tire into water, we can see the air bubbles escaping from the hole in the tire. The air mattress is probably too large to be immersed entirely in water. Instead, we can spray the air mattress with water and listen to the sound of the escaping air.

In an experimental setting, Gick and Holyoak (1983) found that participants had difficulties in the first step – the noticing of the analogical relation:
They gave the participants the *Duncker radiation problem*. The participants had to suggest a solution for destroying a tumor in the stomach of a patient. They were given the information that a strong ray could destroy the tumor, but this ray also destroys the healthy tissue around the tumor. A weaker ray does not destroy the tissue, but it also leaves the tumor intact
Without any further information, only approximately 10 % of the participants could solve the problem (Gick and Holyoak 1983). The actual solution to the problem is *radiosurgery* which is being used to destroy brain tumors today (Tarkan 2003; Goldstein 2011, p.341). It is sketched in Figure 6-2: Many weak gamma rays are focused on the tumor in the patient's head. As they intersect at the tumor, it is destroyed. The brain around the tumor remains unaffected.

If the participants were given one of the inspirations illustrated in Figure 6-2, the portion of participants who could solve *Duncker's Radiation Problem* increased (Goldstein 2011, p.340–346):

- To approximately 30% for the *Fortress Story (Gick and Holyoak 1983)*:
- To approximately 70% on average for the *Lightbulb Problem* in the *Fragile-Glass Condition* respectively (Holyoak and Koh 1987)
- To approximately 30% on average for the *Lightbulb Problem* in the *Insufficient-Intensity Version* (Holyoak and Koh 1987)

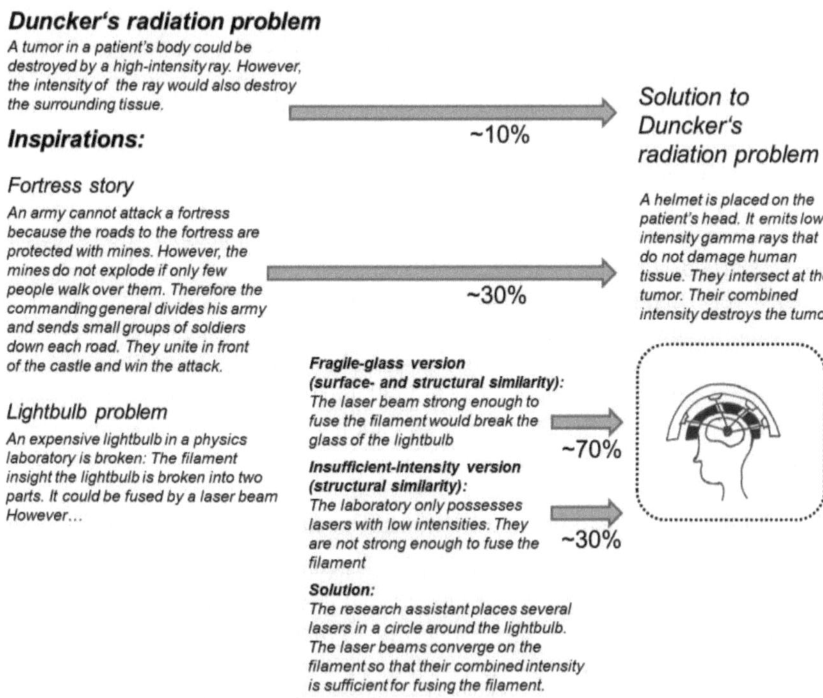

Figure 6-2: Difficulties in noticing analogs - Percentage of study participants solving Duncker's radiation problem (adapted from Gick and Holyoak 1983; Holyoak and Koh 1987; Goldstein 2011, p.340–346)

Why is the effect of each of these inspirations so different?

Cognitive psychologists explain the differences by two aspects of the analogs: *Surface* and *structural* similarity (Goldstein 2011, p.340–346).

If the inspiration has similar elements as the problem to be solved, they are surface similar. For example, the *Lightbulb Problem* has similar elements as *Duncker's Radiation problem*: the laser that can fix the filament is similar to the rays that can destroy the tumor.

If the inspiration shares the underlying challenge or function of the problem to be solved, they are structurally similar. This is the case for the *Lightbulb Problem* in the *Fragile Glass Condition*: the filament can only be fixed by a strong laser but at the same time the glass of the lightbulb would be destroyed by the strong laser beam. The challenge of not destroying the glass is similar to the challenge of not destroying the tissue of the human body in *Duncker's Radiation Problem*.

The *Lightbulb Problem* in the *Insufficient Intensity Version* has the same solution as in the *Fragile Glass Condition*. However, its challenge is described under a different aspect – instead of focusing on the destructive effect of the laser, the single laser beam is described as too weak to fix the filament. The solution is exactly the same as in the Fragile Glass condition, but the inspiration is structurally different from Duncker's Radiation Problem. This decreased the proportion of participants that could solve the problem to less than half. The proportion of participants who could transfer the analogy from the *Insufficient Intensity Version* is similar to the proportion that could transfer the analogy from the *Fortress Story*. To conclude, at least in this experiment, the structural similarity seems to be more important than the surface similarity (Holyoak and Koh 1987).

The experimental approach to studying analogical transfer has been criticized by researchers who observed teams at work. For example, Dunbar and Blanchette (2001) regarded laboratory meetings of biologists and concluded that they used analogies very frequently. Moreover, they transferred both superficial (often for fixing experiments) and structural similarities (often for generating hypotheses). However, these analogies were mostly in-domain analogies. In addition, the biologists did not solve a problem similar to the Duncker's Radiation Problem. Engineering tasks resemble the Duncker's Radiation Problem more than the type of problems solved by the biologists.

6.2 Analogical Transfer in Technical Product Development

In technical product development, analogies have been recognized as important sources of new solutions and design. Analogical design or *Design-by-analogy* implies "*reminding and transferring knowledge of one design situation to another*" for different design tasks in the new situation (Goel 1997). A number of studies have examined the effects of analogical transfer (e.g. Chan et al. 2011 Dahl and Moreau 2002; Srinivasan et al. 2013).

Compared to the cognitive psychologists' approach, the studies focus more on the practical aspects of design-by-analogy, such as the impact of the different sources of inspiration depicted in Figure 4-5 (p. 85). The inspiration sources can be characterized by their type of representation (text, image, video…), their distance to the technical task (in-domain, between-domain) and the comprehensiveness of the information. Figure 6-3 shows the considered factors and the analyzed factors on the analogy and solution level.

On the analogy level, researchers examine if the solution principle of the inspiration is identified and used. Moreover, they determine the level of abstraction of the transfer. It can span from actually using the inspiration for a technical solution to transferring an abstract high-level function. An anecdotal example for using an inspiration directly is the use of a washing machine as a lassi maker. Lassi is an Indian yogurt drink that is made by mixing milk, yogurt, and spices. To produce it in large quantities, usual blenders are too small. An anecdote from India tells of people producing it using washing machines (see the commercial by HSBC 2009). Duncker's Radiation Problem, on the other hand, is an example for the transfer of abstract functions: *split up* and *intersect at the target*.

On the technical solution level, researchers regard creativity measures used to evaluate technical solutions. This includes the quantity of solution ideas generated in a certain period of time (Lopez et al. 2011). Another measure is the variety of the ideas or in other words, the difference between the single ideas (Srinivasan et al. 2013). The quality of ideas can be translated into feasibility and task-specific requirements (Hashemi Farzaneh 2016). The novelty of ideas plays a role, but it is challenging to assess global novelty. Therefore, researchers refer to substitutes, such as the uniqueness of solution ideas in a pool of solution ideas (Wilson et al. 2010).

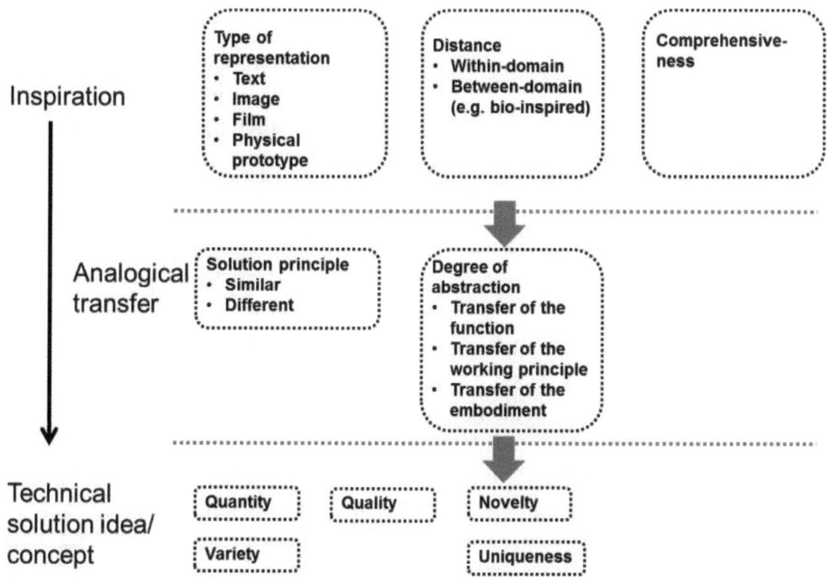

Figure 6-3: Analogies in product development research

The results of research in design-by-analogy are diverse; therefore, we just present a few examples relevant for bio-inspired design. They are all results of design experiments, i.e. the comparison of a group of designers working on technical tasks who were give different inspirational information:

Lopez et al. (2011) regarded the verbal formulation of the inspirations and calculated a semantic distance. For example, "food peelers" are semantically nearer to "nutcrackers" than "depilators". They then analyzed the effect of semantic distance on the **quantity** of solution ideas. They found that the number of solution ideas increased. However, another finding was that the designers evaluated semantically distant inspirations as less useful. The negative attitude of designers towards distant inspirations is a challenge in particular for bio-inspired design.

Keshwani and Chakrabarti (2017) compared the effects of biological inspirations to technical inspirations on different levels of comprehensiveness. They found that biological inspirations explained superficially (low comprehensiveness) inspired more **novel** solution ideas than superficial technical inspirations. However, this was not the case for more detailed inspirations (high comprehensiveness).

Similarly, Wilson et al. (2010) compared biological inspirations to technical inspirations. They found a positive impact of biological inspirations on **uniqueness** used as a measure for novelty of solution ideas. In contrast to technical inspirations, the **variety** of solution ideas did not decrease.

In experiments with pairs of biologists and engineers, we found that the transfer of the solution principle of the biological inspiration resulted in a higher **quality** of solution ideas (Hashemi Farzaneh 2016).

To conclude, several studies indicate positive effects of inter-domain inspirations, such as biological systems. However, the design of the studies differs considerably with regards to technical tasks, inspirational information, and so on. Moreover, the measures to evaluate the solution ideas differ as well.

On the downside, negative effects of design-by-analogy have been observed as well: **design fixation** describes the *"blind adherence to a set of ideas, limiting the output of conceptual design"* (Jansson and Smith 1991). To explain design fixation, they proposed the *Theoretical Model of Conceptual Design* which consists of two mental spaces: the *configuration space* and the *concept space*. The configuration space is comprised of mental representations of physical objects and their configuration, whereas the concept space is comprised of knowledge and understanding. A comparable model of two spaces has been proposed by Hatchuel and Weil (2003) and described as the *C-K theory*. It contains a space of knowledge (K-space) and a space of concepts (C-space). Unfortunately, the naming is inverse to Jansson and Smith's model (see Figure 6-4).

According to Jansson and Smith (1991), designs require movement between the two spaces, and design fixation can be regarded as a mental block of these movements. This is illustrated in Figure 6-4. If we consider the torrent frog example, the toe pad microstructure seems to be relevant for the torrent frogs' adhesion capabilities. If we transfer this knowledge to generate ideas for the design of a climbing glove, we move from the concept space (K-space) to the configuration space (C-space). This step is a step of *realization* (Kroll et al. 2001, p.62) or *disjunction* (Hatchuel and Weil 2003). Within the configuration space (C-space), we can refine, choose, structure, and expand this solution idea. For example we can generate variations of different patch area and patch configuration. However, without moving back to the concept space (K-space), no genuinely new concepts can be generated. This is when design fixation occurs. One reason to move back to the concept space is to conduct experimentation. This entails a step of *abstraction* (Kroll et al. 2001, p.62) or *conjunction* (Hatchuel and Weil 2003). For example, the torrent frog's hexamer microstructure may not be able to be reproduced accurately using artificial material. Still, experiments may show that artificial round microstructures enable a sufficient adhesion. This new knowledge enables the generation of a new solution idea, gloves with patches with a round microstructure.

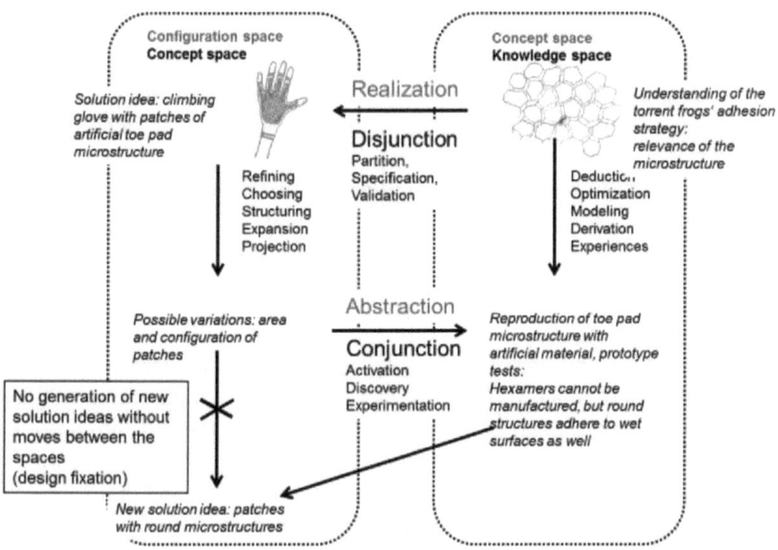

Figure 6-4: Design fixation explained via the Theoretical Model of Conceptual Design (grey letters), and the C-K Theory (black letters) (Jansson and Smith 1991; Kroll et al. 2001; Hatchuel and Weil 2003; Hatchuel et al. 2011)

Design fixation has been reported in relation to the presentation of inspirations such as existing products or biological systems. In particular, the presentation of images seems to induce fixation (Chrysikou and Weisberg 2005; Cardoso and Badke-Schaub 2009). To mitigate design fixation, the following measures to be taken are proposed:

- Increasing awareness of designers:

  Several researchers have observed that designers who are aware of the phenomenon of design fixation can avoid it more easily. For example, Chrysikou and Weisberg (2005) gave designers inspirational examples, but instructed them successfully not to use certain elements for their own solutions. Hatchuel et al. (2011) taught students to consciously move between the spaces illustrated in Figure 6-4 to reduce design fixation. Experience can increase awareness, for example experienced designers comment that an initial idea can block subsequent ideas (Crilly 2015).

- Using physical prototyping:
  In design experiments, the use of physical prototypes decreased design fixation and increased creativity (Youmans 2011)
- Providing incubation time
  Incubation time allows designers to take a step back from the design task and the developed solutions. Getting back to work, they are more likely to consider new aspects and ideas (Smith and Linsey 2011).

To conclude, so far, we have considered mainly unintentional fixation effects. However, designers can also intentionally fixate on particular ideas. In particular, experienced designers sometimes prefer familiar solutions that "worked well" in the past (Youmans and Arciszewski 2014). This is not necessarily a bad thing, but it excludes potentially better new unfamiliar solutions, such as bio-inspired solutions.

6.3 Methods for the Transfer of Bio-inspired Analogies

Figure 6-5 shows an adaptation of the general view on analogical transfer (see Figure 6-1, p. 142) to bio-inspired design. The two regarded domains are the biological and technical domain. If we follow a biology push approach, we focus on the tree or torrent frog's toe pads first. An analysis of the toe pads provides the insight that the toe pads, among other functions, absorb and emit fluid. If we search for this abstracted problem, we might find out that irrigation and drainage systems must solve a similar problem. We can transfer an analogy from the toe pads to develop an innovative drainage system: Hexamer elements provide space for plants. The channels between the hexamer elements drain excess water. Putting pressure on the elements squeezes water out of the channels to irrigate the plants.

If we take a technology pull approach, we start by analyzing the technical task or problem. We can formulate an abstracted problem and search for adequate biological inspirations. We might discover the tree or torrent frog and its toe pads as potential analogs for developing an innovative drainage system.

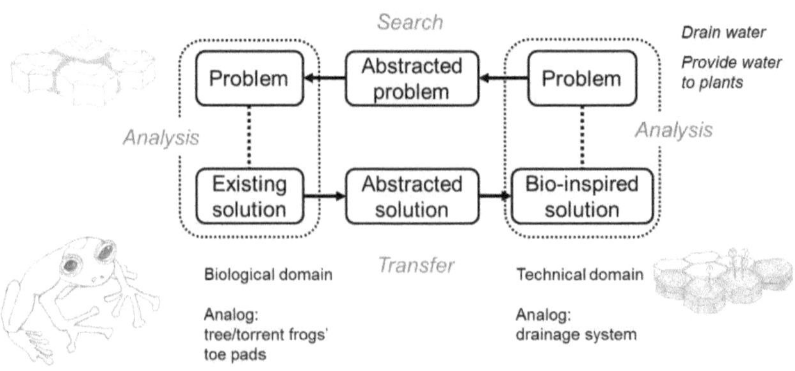

Figure 6-5: Analogical transfer in bio-inspired design

To conclude, the transfer of bio-inspired analogies is independent of the starting point of the design process (biology push or technology pull). It requires a step of abstraction, as biological systems can rarely be used in the technical domain directly.

The six methods presented in this chapter support the formulation of different analogy types. As shown in Figure 6-6, they focus on different aspects:

The *Analogy Model* and the *Causal Relation Template* support the identification of the relations between the elements of the biological and technical systems. An example would be the relation between the torrent frog's adhesion capacities, the microstructure of its toe pads, and the fluid it produces. The *Causal Relation Template* focuses on the formulation of (simplified) causal relations. Therefore, it is particularly suitable for creative ideation. The *Analogy Model* takes a more holistic viewpoint and can therefore be used in all bio-inspired design scenarios.

The *Decision Model Abstraction Level* and the *Formulation of Different Analogy Types* focus on the transfer of analogies to different abstraction levels. The *Decision Model Abstraction Level* suggests a decision procedure for finding an adequate analogy. The *Formulation of Different Analogy Types* proposes analogy types used in bio-inspired design research. They can both be used in all bio-inspired design scenarios. However, the *Formulation of Different Analogy Types* gives very clear instructions which are particularly useful for creative ideation. The two methods can also be combined; if we follow the process suggested by the decision model, we can use the instructions to formulate different analogy types.

BioId and *Synectics* are particularly useful in supporting creative ideation in teams. *BioId* consists of a template which supports the transfer of a concrete and an abstract analogy. *Synectics* is a creativity method that has been developed for professional interdisciplinary design teams. It is not restricted to bio-inspired design but supports a wide range of inspirations, in particular unobvious inspirations. It is included in this book because it supports bio-inspirations, and, more importantly, because it suggests a complete procedure for creative ideation in teams of designers with different disciplinary backgrounds.

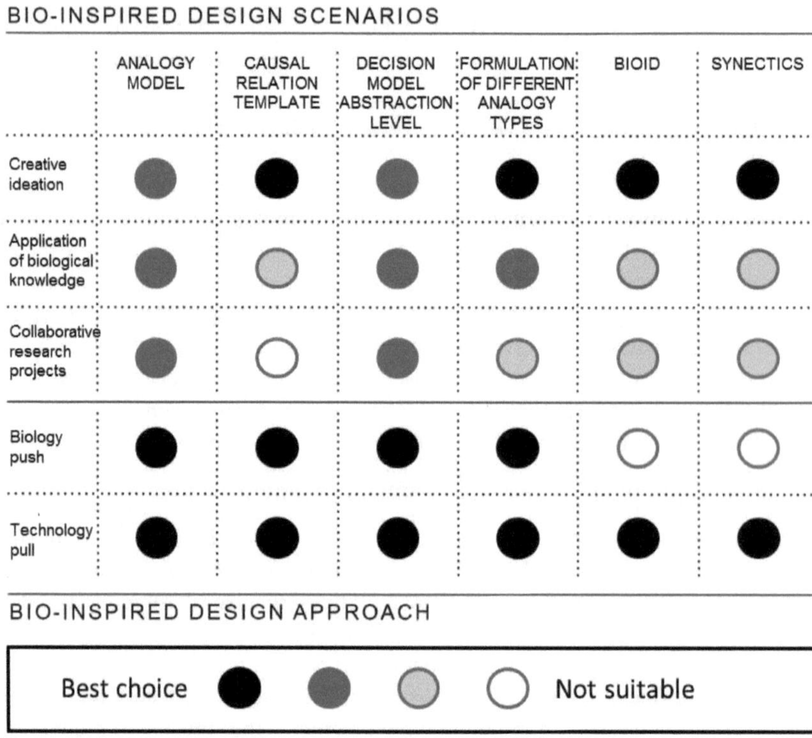

Figure 6-6: Suitability of transfer methods for different bio-inspired design scenarios and approaches

BEST PRACTICE

Stair-climbing Wheelchair – Analogical Transfer Inspired by Multiple Biological Species

University of Bayreuth, Technical University of Munich, Bosch und Siemens Hausgeräte GmbH, Sandler AG

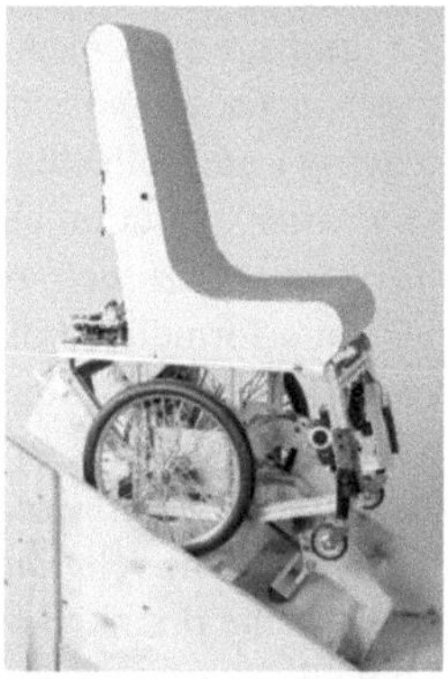

Photo: Uli Benz/TUM

Analogies can be based on multiple biological species and occur on very abstract levels. The stair-climbing wheelchair is a prime example. The project's aim was to develop a technical device which enables wheel chair users to independently climb stairs. Current solutions either require the installation of devices on the stairs (e.g. stair lifts) or other persons must help the wheel chair users. To develop the stair-climbing wheelchair, designers considered evolution: when water animals first moved out of the seas to conquer the land, they did not have feet to walk. Instead, when they met an obstacle, they propped themselves up on their extremities and crawled over it. This "propping up" movement served as an inspiration to develop two supports that prop up the wheel chair and lift it on the next step. When the user wants to descend, they support the wheel chair to prevent it from rolling down the stairs. The two supports are folded under the wheelchair when it is used on plain surfaces. The developed wheelchair has only one axis and stabilizes itself as a pendulum. This leaves enough space for the two supports and makes the wheelchair more agile and easily maneuverable.

Based on an interview with Professor B. Wolf (July 2017), and Technical University of Munich (2016)

6.3.1 Analogy Model

Fact Sheet p. 211

The *Analogy Model* suggested by VDI 6220, 2012 (2012) proposes a three-step procedure for transferring analogies from a biological to a technical system.

The first step is the collection of aspects of the biological system and the technical system. Aspects include all relevant information on the systems, such as function/purpose, properties, embodiment, performance, and so on. There is no restriction as to what can be considered as an aspect. The aspects are not structured or put into an order. Figure 6-7 shows an example – an excerpt of an *analogy model* of the torrent frog and a drainage system. As the figure shows, several functional aspects (e.g. *adhere to wet surfaces*, *emission of fluid* etc.) and several structural aspects (e.g. *micro-hexamers*) are collected for the frog. For the drainage, functional aspects (e.g. *take-up water*), solution elements (*channels*) and additional requirements (e.g. *space for plants and their roots*) are listed. The aspects are unstructured and similar aspects are formulated differently (e.g. absorption of fluid, take-up of water). Moreover, the analogy model of the drainage system includes both a common solution, i.e. *channels* to drain water and a new desired function that this solution does not fulfil (*emit water*).

In the second step, for each system, the relations between the aspects are established. For example, in Figure 6-7, the drainage system's functional aspects *take-up*, *store* and *emit water* are linked. The requirements *space for plants* and *potting soil* are not directly linked to these functional aspects.

In the third step, the relevant aspects are merged: similar aspects are identified and all relevant aspects connected to them are integrated into a model of the bio-inspired system. The procedure is a variation of the general analogical process shown in Figure 6-5: the general analogical process is based on the assumption that similarities are identified between the *problems* or tasks of biological and technical system (horizontal relation). Then, the solution of the analog is analyzed (vertical relation) and transferable aspects are mapped to the target system (horizontal relation). In the case of the *Analogy Model*, it is not explicit "problems" that are mapped, but rather all similar aspects that can be mapped. For example, in Figure 6-7, the *channels* of the technical drainage system are considered similar and mapped to the *spaces between the hexamers*. Then, all irrelevant aspects are deleted and new relations are established. In our example, this results in a bio-inspired drainage concept: hexamer elements provide space for potting soil, plants, and their roots. The channels between the elements take-up and store water. Pressure on the elements leads to an emission of previously stored water.

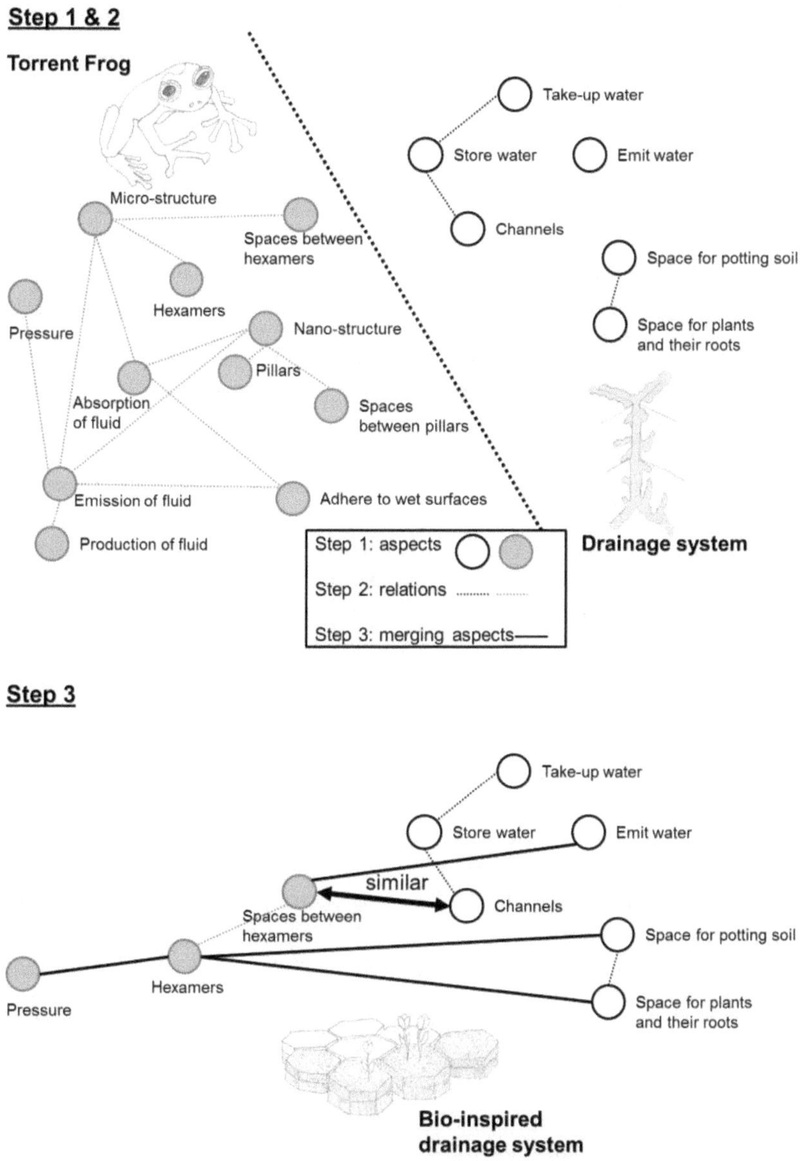

Figure 6-7: Analogy Model example

The *Analogy Model* supports a structured reflection on differences and similarities between technical and biological systems. It visualizes the relevant aspects and their relations. As it does not impose strict definitions for these aspects, it can be used very flexibly and without much effort to understand concepts such as function, behavior, and structure. On the downside, designers are left without guidance to collect the relevant aspects of biological and technical system.

The *Analogy Model* can be used for biology push and technology pull approaches and for all bio-inspired design scenarios equally. Moreover, the method is suitable for different stages within a bio-inspired design project.

One suitable use is after the analysis of biological and technical systems: by visualizing and structuring information, new ideas for transfer can emerge. Moreover, the *Analogy Model* can be used at the start of a bio-inspired design project to collect and structure all available information of biological and technical systems. Then, it highlights missing information and initial ideas to provide starting points for the analysis phase.

6.3.2 Causal Relation Template

Fact sheet p. 213

In student classes on bio-inspired design, researchers frequently observed that students have difficulties in mapping biological solutions to technical tasks or problems (Shu and Cheong 2014). In particular, they found that students focused on mapping similar structural elements instead of relations (Cheong et al. 2010). This initiated the development and testing of the Causal Relation Template. This template demands the formulation of a biological solution using two functions: the *enabling function* and the *desired function*. One desired function for the torrent frog and its adhesion strategies is to *enlarge the contact zone* between itself and the rock. Which function(s) enable the desired function? On an integral level, the frog presses its entire body to the rock. Furthermore, on the micro- and nano-levels, the toe pads' skin structure, i.e. the micro-hexamers and nano-pillars, are deformed by the frog's weight. Both strategies can be formulated and mapped to a technical task using the *Causal Relation Template*:

1. A sentence describing the torrent frog's strategy on an integral level can therefore be formulated as:

"Torrent frog presses itself closely (enabling function) to rock to enlarge (desired function) contact zone."

This strategy can be mapped to a climbing glove, for example by formulating:

"Climbing glove adapts (enabling function) to shape of rock to enlarge (desired function) contact zone."

2. On the micro- and nano-level, the torrent frog's strategy can be described as:

"Torrent frog's weight deforms (enabling function) toe pads' micro- and nanostructure to enlarge (desired function) contact zone."

This strategy can be mapped to a climbing glove, for example by formulating:

"Climber's weight deforms (enabling function) glove's micro- and nanostructure to enlarge (desired function) contact zone."

In the next step, designers can concretize the technical solutions that implement the enabling function: how could the gloves adapt their shape to the shape of the rocks? How can the glove's material be structured on the micro- or nano-level to deform under pressure?

Using the *Causal Relation Template*, designers can quickly identify causal relations of a biological system and reflect on possible analogical transfer to a bio-inspired technical solution. However, it does not support the identification and transfer of more complex multi-step relations. For example, the torrent frog's adhesion is achieved by a combination of multiple principles: the micro- and nanostructures also absorb fluid in wet environments and emit fluid in dry environments, which causes a capillary effect, friction, and more. The Causal Relation Template can be used to formulate each causal relation separately, but it cannot represent the complex relations between the single principles. In addition, even though the *Causal Relation Template* aims at reducing fixation on the structural elements of a biological system, it could possibly induce fixation on a single causal relation (Cheong et al. 2010).

To summarize, the *Causal Relation Template* is suitable both for biology push and technology pull approaches. Due to the low effort it requires for use, it is most appropriate for creative ideation. To a certain extent, it can be used for the application of biological knowledge. With increasing knowledge and complexity of the regarded relations, the method becomes less useful.

6.3.3 Decision Model Abstraction Level

Fact sheet p. 215

In bio-inspired design projects with undergraduate students, Gramann (2004, p.97–104) developed a procedure which can be used for deciding on the appropriate abstraction level. Similar to procedural models presented in chapter 2, Gramann (2004) suggests formulating the intention or target of the bio-inspired design. Regarding the climbing gloves example, the target could be to improve the gloves' adhesion to wet rocks. In the next step, the target formulation can be used to perform a search as described in chapter 4. Designers could, for example, discover the torrent frog's adhesion abilities. An analysis of the torrent frog (literature / consultation of experts / experiments etc.) must be conducted in order to understand which properties of the torrent frog's toe pads contribute to adhesion. To summarize, these three steps are typical bio-inspired design stages that were described in the previous chapter. However, the *Decision Model Abstraction Level* then proposes three questions and iterations (Gramann 2004, p.97–104; Lindemann and Gramann 2004):

Firstly, the designers must attempt to transfer the biological solution to a technical solution. Figure 6-8 shows an example regarding the torrent frog: one option is to mimic both micro- and nanostructures with an artificial material that can be used for the glove surface. However, even if it can be manufactured, this solution may be very expensive.

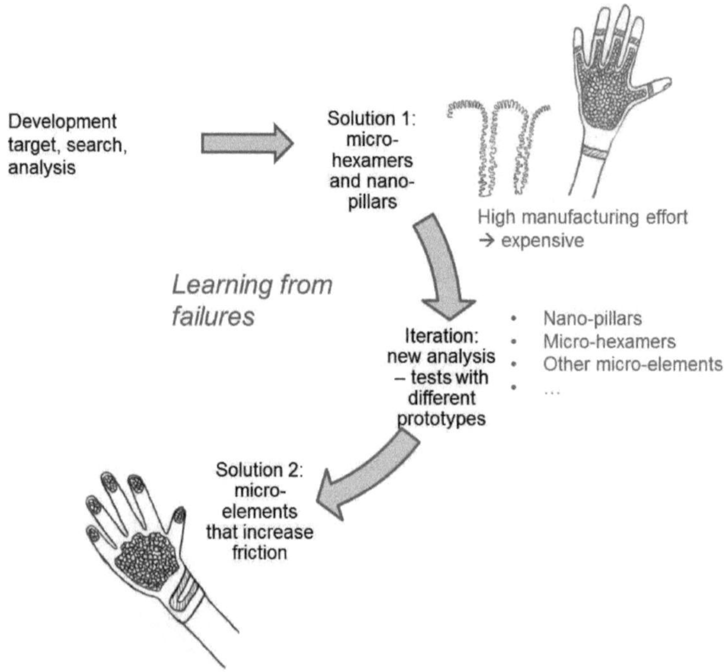

Figure 6-8: Learning from failures – sample iterative process using the Decision Model Abstraction Level

Therefore, secondly, the designers must ask the question of whether the abstraction level is adequate. In the case of the climbing gloves, the abstraction level is probably too concrete. As a consequence, Gramann (2004) proposes an iteration, i.e. designers should go back to the analysis stage. In the torrent frog example, further analyses could provide insights into the most important properties or physical relations of the torrent frog's toe pads for their adhesion. At this point, we must take into account that designers probably have also gained additional understanding of the adhesion phenomenon by developing, building, and possibly testing the first solution. In technical product development, researchers have named this *co-evolution of problem-solution* (Cross 2001). Even though they refer to the further development of the technical problem or task, the deeper understanding of the source of anlogy – the biological system – can equally co-evolve with the bio-inspired solution. In our hypothetical example, designers therefore conduct experiments with the first overly expensive solution prototype. Based on these tests, they build further simple prototypes that mimic aspects of the toe pad structure on different levels of abstraction. Examples would include one prototype with nano-elements similar to the pillars, one prototype with micro-hexamers only, or one prototype with different micro-elements. The analysis of these tests makes the designers understand that it suffices to put micro-elements on the gloves which deform under pressure and increase friction. This knowledge enables the development of the final solution.

If the change of the abstraction level does not result in a better bio-inspired solution (idea), Gramann (2004) suggests a third question; designers should question the feasibility of the development target. If it still seems feasible, he suggests searching for other biological systems. If it is does not, he proposes to redefine the development target.

To summarize, the *Decision Model Abstraction Level* has several advantages. It demands a critical reflection of the decisions that must be taken during the analogical transfer. Moreover, it implies learning from failures, a very constructive approach to improve bio-inspired solutions that do not fulfil the requirements. A disadvantage of this method is that it does not provide criteria for evaluating the bio-inspired analogy; designers might tend to make decisions too hastily on the appropriateness of the abstraction level or the feasibility of the development target.

The *Decision Model Abstraction Level* can be used for biology push and technology pull approaches and for all bio-inspired design scenarios.

6.3.4 Formulation of Different Analogy Types

Fact sheet p. 217

Strictly speaking, the Formulation of different Analogy Types is not a method. Researchers have used different definitions of analogy types to classify bio-inspired analogies. We propose to use these definitions for generating several analogies based on a single biological inspiration. They can stimulate different ideas for analogies. We present the definitions of accuracy, similarity, and elements of transfer. An overview is shown in

Table 6-1. The comparison is based on the four aspects *function* (i.e. purpose of the biological or technical system), the *physical effect* (natural law), the *working principle* (material and geometric properties), and the *direct use* of biological system elements.

To start with ***accuracy***, Mak and Shu (2004b) and Vakili et al. (2007) define four categories. <u>*Unrelated strategies*</u> are not linked to the biological system. An example is putting glue on climbing gloves to improve adhesion. If we suppose that the torrent frog served as an inspiration, this strategy is unrelated. <u>*Incorrect strategies*</u> are inspired by the biological system, but the designers misinterpret the biological *strategy*. In other words, the bio-inspired solution has similar geometric or material properties on the working principle level, but does not implement the physical effect of the biological system. An example is when the torrent frog's adhesion strategy is confused with the gecko's. The result might be a climbing glove coated with fine hair similar to the gecko foot. <u>*Incomplete strategies*</u> use a "general principle" behind the biological strategy, but not the concrete strategy (Vakili et al. 2007). They implement only a function similar to the biological inspiration, but have a different physical effect. For example, designing a climbing glove with a rough leather material would implement the general function *increase friction*, but not a more concrete solution from the torrent frog. Correct strategies use "similar elements" as the biological system (Mak and Shu 2004b). *Correct strategies* refer to bio-inspired solutions which implement a function and the associated physical effect of the biological inspiration. They can also implement the working principle and embodiment of the biological inspiration. An example would be to design a glove with a micro- or nanostructure similar to the torrent frog's toe pad structure.

Regarding ***similarity***, Mak and Shu (2004a) define two dimensions: the *strategic accuracy* and the *abstraction of biological entities*. The four resulting categories are illustrated on the fact sheet. An analogy is considered *strategically accurate* if the biological strategy to solve the addressed problem is transferred to the technical solution idea. The biological entities are considered *abstracted* if no elements from the biological inspiration are directly used in the technical solution idea.

The categories *biological transfer* and *literal implementation* both refer to the direct use of the biological system (or its elements). A hypothetical example involves using frog secretion: in the case of a *biological transfer*, frog secretion is used to moisturize the climber's hands. This analogy is not similar in terms of physical effect or function. Unlike biological transfer, *literal implementation* additionally requires a similar physical effect and function. For example, if we use frog secretion to increase viscosity and friction for a better adhesion on dry, smooth surfaces, this can be considered a literal implementation.

The categories *analogy* and *anomaly* refer to a bio-inspired solution that does not directly use elements of the biological system. An *analogy* is transferred if there is no similarity on the embodiment level, but at least on the function level, and possibly on the working principle and physical effect level. This would be the case for leather gloves or gloves with a micro- or nanostructure similar to the torrent frog. The bio-inspired solution is an *anomaly* if it has no similarity with the biological solution or a similarity exclusively on the working principle or function level. Designing gloves with suction cups is one example.

The ***elements of transfer*** (Sartori et al. 2010) are based on elements of the SAPPhIRE model (Chakrabarti et al. 2005). Sartori et al. (2010) analysed bio-inspired solutions and identified five abstraction levels of bio-inspired transfer: *Copy parts* refers to the direct use of the biological material and structure for the technical solution, such as using frog secretion for any bio-inspired solution (hand moisturizer, increasing viscose friction, etc.). *Transfer organs* implies transferring properties or conditions that are necessary for a function; i.e. the bio-inspired system must use at least the physical effect and the function of the biological system. Gloves with a micro- or nanostructure similar to the torrent frog's toe pads are an example. *Transfer attributes* entails the transfer of properties (working principles), but it is unclear if they contribute to fulfilling the addressed function. In other words, the physical effect is unclear or not similar to the biological system. If we design gloves with a macro-hexamer structure, the following would be the case:

The change of dimensions can result in lower adhesion forces as the capillary effect decreases. *Transfer state change* describes a more abstract analogical transfer; that is, the technical system addresses a similar function to the biological system, but not with the same means. Leather gloves that increase friction are an example for the transfer of a state change. *Resulting transfer* implies an unintended transfer of analogies or similarities from the biological system. This category is disregarded here because it cannot be used as instruction for analogical transfer.

Table 6-1: Analogy types (adapted from Hashemi Farzaneh 2016, p.45)

Analogy types		**Comparison to the biological inspiration (x: different, o: different or similar, ✓: similar)**			
		Direct use of biological system (elements)	Working principle*	Physical effect*	Function*
Accuracy (Mak and Shu 2004b)	Unrelated	x	x	x	x
	Incorrect	o	✓	x	o
	Incomplete	x	x	x	✓
	Correct	o	o	✓	✓
Similarity (Mak and Shu 2004a)	Biological transfer	✓	✓	x	x
	Literal implementation	✓	✓	✓	✓
	Analogy	x	o	o	✓
	Anomaly	x	o	x	o
Elements of transfer (Sartori et al. 2010)	Copy parts	✓	✓	o	o
	Transfer organs	x	o	✓	✓
	Transfer attributes	x	✓	x	o
	Transfer state change	x	x	x	✓

*working principle: combination of geometric and material properties; physical effect: physical law that supports the function; function: the purpose of a technical/biological system and its elements

As Table 6-1 shows, the definitions of accuracy, similarity, and elements of transfer are not clearly distinguished, they partially overlap. Table 6-2 shows the categorization of the discussed bio-inspired climbing glove examples into the categories accuracy, similarity, and elements of transfer.

Table 6-2: Analogy types: examples for the overlap between accuracy, similarity, and elements of transfer

Accuracy	Similarity	Elements of transfer	Example
Unrelated	Anomaly	No transfer	Gloves with fine hair (similar to gecko)
Incorrect	Anomaly	Attributes	Gloves with suction cups
	Biological transfer	Parts	Frog secretion in moisturizing hand lotion
Incomplete	Analogy	Attributes	Gloves with macro-hexamer structure
		State change	Leather gloves
Correct	Literal implementation	Parts	Frog secretion to improve viscous friction on smooth surfaces
	Analogy	Organs	Gloves with micro- or nanostructure similar to the torrent frog's toe pad structure

To effectively formulate different analogy types, single designers or better a team of designers can regard the definitions and examples for the categories *accuracy*, *similarity*, and *elements of transfer*. The definitions and examples serve as stimuli to generate different analogies. Moreover, they promote a reflection on the correctness and degree of abstraction of analogies. However, the definitions should not be taken too seriously for ideation purposes, as designers could exclude potentially good solutions because they categorize them as *incorrect*, *anomalies*, or otherwise. If working principles or physical effects of the biological system are not transferred, this does not mean the analogy is inappropriate for a technical solution. Even though the generation of feasible, high-quality solutions based on such analogies seems to be more random, they must be considered for further development.

The *Formulation of different Analogy Types* can be used for biology push and technology pull approaches. As it builds on creativity, the method is most suitable for creative ideation. It is also appropriate for the application of biological knowledge. It seems, however, less useful for collaborative research projects, except for workshops, which can provide new ideas for a different analogical transfer.

For the elements of transfer of the type *analogy*, the SAPPhIRE method can be used to abstract the biological or technical system.

BEST PRACTICE

Hexal LED Trunking System – Application of Vault Structures in Multiple Technical Applications

Osram GmbH, SITECO GmbH

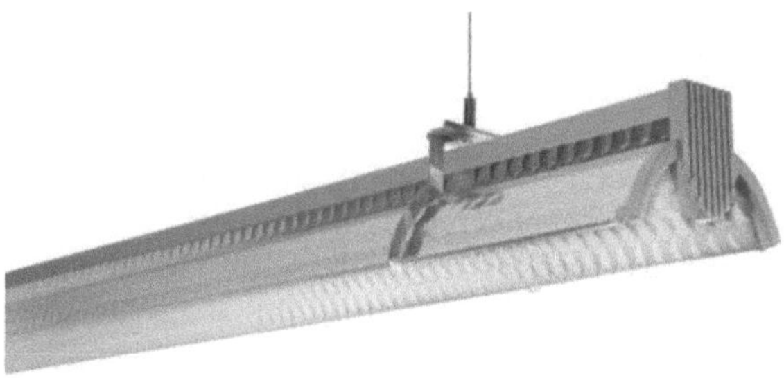

Trunking Luminaire SITECO HEXAL LED, Photo: Osram GmbH

Biological inspirations such as vault-structures can be applied in completely different technical systems. At first glance, the increased mechanical stability of vault structures make them interesting for lightweight design of components that have to resist high forces. However, biological solutions have multiple properties, some of which are only discovered upon closer look. This is the case for vault structures: Apart from increasing mechanical stability, they diffuse light and can generate a comfortable glare-free lighting situation. This property inspired the Hexal LED trunking system. Here, vault structured sheets were used to integrate housing and reflector in a single component. Since the vault-structured sheets are very thin, the Hexal LED trunking system reduces weight up to 80% in comparison to other LED trunking systems. The single-component design facilitates the installation of the trunking system.

Based on an interview with Osram designer R. Wrenger (June 2017)

6.3.5 Biold

Fact sheet p. 219

BioId is a method specifically designed for creative ideation in teams of biologists and engineers (Hashemi Farzaneh 2016, p.74–89). It addresses the different working approaches and language barriers between biologists and engineers by suggesting they sketch and use familiar abstractions. Another aim is to prevent fixation on a single biological inspiration or analogy during an ideation workshop. The detailed aims of the method are:

- To support a structured procedure to use several biological inspirations
- To support sketching to increase the understanding of biological systems
- To support the transfer of correct analogies over two abstraction levels
- To support the discussion of the task at the beginning of ideation

To achieve these goals, *BioId* was conceptualized using elements from the *KoMBi* and *Biocards* abstraction methods (Hashemi Farzaneh 2016). The resulting concept is illustrated in Figure 6-9.

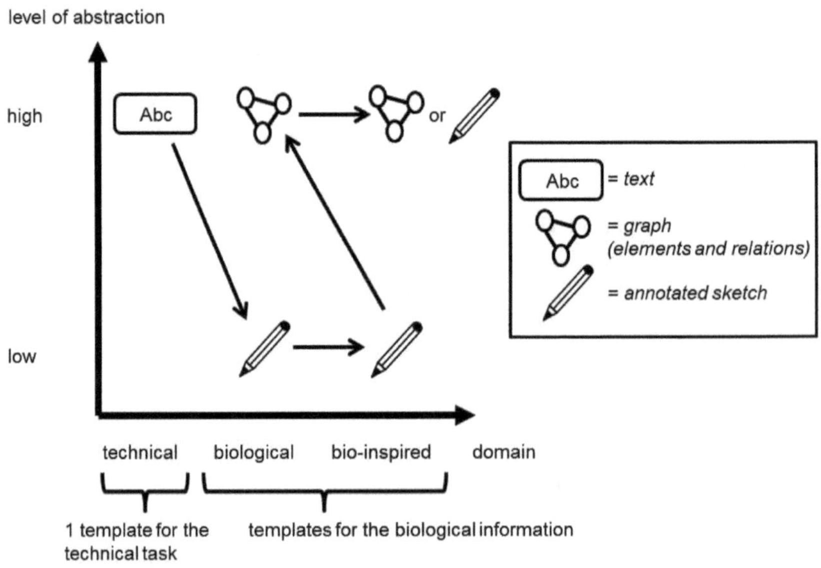

Figure 6-9: *BioId* – procedure and means of representation (Hashemi Farzaneh 2016)

To start with, designers should define the main task and desired properties of the technical solution to be developed. These questions aim to foster a discussion about the technical task and a common understanding of biologists and engineers. In the climbing glove example, the task could be to design gloves with improved adherence to wet surfaces. Relevant properties are, for example, the hand size, the weight to be supported, and so on.

In the next step, designers must consider a biological system. As shown in Figure 6-9, they start with the concrete embodiment level. They are asked to sketch the elements of the biological system that they consider relevant for the technical task and its desired properties. In the case of the torrent frog, this could be the micro-hexamer structure. Based on this sketch, they have to transfer the first analogy and then sketch it. Gloves with a micro-hexamer structure are an example of such a concrete analogy.

Then, designers should turn to a more abstract level and use a simple graph representation to explain the relations of the biological system's elements (see Figure 6-9). One of the physical effects implemented by the micro-hexamers is friction, in that the frog puts weight on the micro-hexamer structure. Consequently, the hexamers deform and the resultant increased contact area between hexamers and the surface increases friction. This abstract relation can be transferred to a glove. However, the micro-hexamer structure does not have to be mimicked. Other shapes with different size can be used as well. Therefore, at this point, a more abstract analogy is transferred to develop gloves with elastic structures or elements. These elements could be, for example, macro-sized nubs. Designers are left free to provide a graph representation of their solution idea or to sketch it.
The abstract analogy is comparable to the analogical transfer requested by the *Causal Relation Template* (see section 6.3.2, p. 154 and fact sheet p. 213). However, the graph representation allows more complex relations to be modeled than the verbal description of the causal relation does.
The advantages of *BioId* is the support of a structured and efficient approach in ideation workshops. Using the template designers can consider several biological systems and are advised to transfer at least two analogies per biological system. Visualization (sketches, graph representations) supports the interdisciplinary communication between biologists and engineers. On the downside, the use of the *BioId* template requires time and effort. Moreover, designers can lose motivation if they feel unable to represent the biological system on a concrete and abstract level.
BioId is based on a technical starting point and is therefore appropriate for technology pull approaches. As it is designed for creative ideation, it is less suitable for the application of biological knowledge and collaborative research projects.

6.3.6 Synectics

Fact sheet p. 221

The *Synectics* method (Gordon 1961) is a technique for creativity workshops with interdisciplinary groups. Based on studies of creativity conducted in the 1940s to 1960s, it proposes a very unconventional approach to creative problem solving. It has a high potential to be applied in bio-inspired design because it builds on interdisciplinary groups and works with analogies and metaphors.

To understand and apply *Synectics*, it is helpful to understand its research background. Through ideation sessions and interviews, Gordon (1961, p.18–19) identified four psychological states that played a role for ideation:

- Detachment and involvement: designers feel alternating states of closeness or distance to the technical system or task (*How would I feel if I was this screw? ↔ I have to take a step back for an overview*)
- Deferment: designers consciously resist fixation on a solution idea (*We really have to move on, otherwise we will just develop the same solution all over again!"*)
- Speculation: designers break the rules of natural laws or consciously ignore the constraints of the technical task (*What if we could just "beam" this object to ...?*)
- Autonomy of object: designers have the impression that a good solution develops itself, they perceive a flow of ideas *(Wow, I feel that this solution is evolving itself, without me influencing it!*)

Studies with groups working on ideation tasks showed that "*making the familiar strange*" could be used as a strategy to achieve these four psychological states (Gordon 1961, p.28–31). Based on this observation, a structured procedure for groups was developed. A crucial aspect of *Synectics* is the selection of the group members. Gordon (1961, p.12–13) proposes two criteria:

1. The group should be diverse regarding domain expertise, i.e. it should be interdisciplinary
2. The group should be diverse regarding personality ("*emotional constitution*").

According to Gordon (1961), the diversity of personalities is more important than the diversity of disciplinary background. However, relevant personality traits can be difficult to evaluate in practice. Aspects that can be considered include whether a potential group member quickly takes decisions or evaluates arguments carefully, is persistent or rather passive, and can accept criticism.

To ensure feasibility of the solutions developed in *Synectics* groups, an expert from the application area should participate in the group. The expert can provide knowledge but also judge on the usefulness of the developed solutions (Gordon 1961).

Synectics groups are typically trained on example problems for a while and collaborate in the long term. Nevertheless, Gordon (1961) emphasizes that they can already begin developing successful solutions in the training phase. Therefore, it is worth conducting *Synectics* on an untrained group as well.

The *Synectics* procedure starts with a technical problem. To *make the strange familiar*, the group analyzes the problem in detail. The outcome is an understanding of the problem, similar to other creativity techniques. Then, the operational mechanisms come into play, where the designers of the *Synectics* group formulate four different types of analogies or metaphors to *make the familiar strange*. The four types of analogies place the designers into the psychological states of detachment/involvement, deferment, speculation, and autonomy of object (Gordon 1961, p.36–56):

To start with, **direct analogies** emerge from the comparison of the technical problem to similar problems which are considered *"parallel"* to the technical context. This is the case for a typical bio-inspired analogy. Regarding an artificial drainage system, designers can ask: *How does nature absorb and emit water? Which biological systems can be regarded?* Possible answers could include the torrent frog or general amphibian feet, but also other biological systems, such as plants, the human metabolism, etc.

To generate **personal analogies**, the designers take the role of the technical systems, one of its components, or the material to be processed. Taking the role of a technical component or a material enables the group members to feel the relevant influences within the technical system. If we take the drainage system as an example, group members could take the role of a water droplet. A question to identify with the water droplet could be: *How do I feel as a water droplet falling into a drainage dig – how do I get out of it again?*

Symbolic analogies originate from illustrative, esthetically satisfying metaphors. Designers search for such metaphors to detach from the familiar, rational view of the technical problem. This supports the generation of novel, uncommon ideas. Regarding the drainage system, designers could wonder: *What takes in liquid and releases it again?*. A possible metaphor could be a sponge, an elephant trunk, a syringe... These metaphors that have little in common with the actual technical problem cause designers change their perspective and thereby initiate uncommon ideas.

Fantasy analogies are based on wishful thinking used in a constructive way. To generate this type of analogy, designers ask: *How do we in our wildest fantasies desire the technical system to work?* The answers can be very abstract such as little dwarfs collecting water in upside-down umbrellas and dispersing the water with watering cans. However, explaining the procedure of the fantasy analogy in detail supports the generation of feasible ideas. For example, upside-down umbrellas collect water due to their concave shape. To get into a watering can, the water could flow through a hole in the ground with this concave shape, and so on.

Using the *Synectics* method, designers are likely to generate concrete bio-inspired analogies as direct analogies. However, the other operational mechanisms can result in more abstract bio-inspired analogies, especially if there are biologists in the group or if biological knowledge has been discussed before.

During ideation, the *Synectics* group documents ideas and, in the next step, selects the most relevant analogies. A comparison with the problem as understood provides arguments as to whether to develop a technical solution or not. At this point, physical prototypes are an important means to the properties of the initial technical solution. The result of the *Synectics* method is a technical solution or a research target (Gordon 1961, p.158–160).

To conclude, *Synectics* supports the transfer of unobvious, rather distant analogies. As an explicit discussion of the technical task is part of the procedure, it can also improve the task definition. On the downside, the preparation and implementation of *Synectics* requires considerable effort due to the selection of designers to participate in a *Synectics* group, learning the multi-step Synectics method, and the relatively long duration of a *Synectics* workshop (at least half a day). Furthermore, designers need to have a certain open-mindedness to be willing to engage in generating *irrational* analogies.

Synectics is designed for technology pull approaches and for creative ideation. For the application of biological knowledge or collaborative research projects it can be appropriate at the very beginning of the project to collect initial ideas.

7 Fact sheets

H. Hashemi Farzaneh and U. Lindemann, *A Practical Guide to Bio-inspired Design*, https://doi.org/10.1007/978-3-662-57684-7_7

PLANNING

Technology Pull Procedure [Lenau et al. 2010]

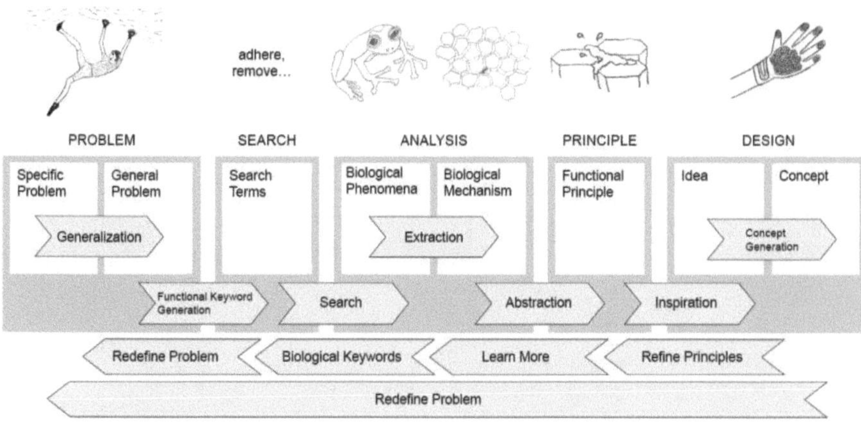

Description

The method describes five consecutive phases and shows reasons for iterations: In the problem definition phase, a specific (technical) problem is generalized. For the search phase, the method emphasizes the importance of formulating search terms. The analysis phase aims at identifying the underlying mechanisms of biological phenomena. In the principle phase, an abstract bio-inspired principle is developed to be detailed in the design phase.

Advantages	Challenges
Practical details on the first phases of product development	Few details on the later phases of product development
Highlights reasons for iterations	

Suitable for...	Biology push ☒	Technology pull ☑

Creative ideation	● ● ○
Application of biological knowledge	● ● ●
Collaborative research projects	● ● ○

Combines best with...	Biocards for ABSTRACTION

PLANNING

Biology Push Procedure [Helms et al. 2009, Vattam et al. 2009]

Biological system	→	Abstraction	→	Technical problem	→	Bio-inspired solution
• Identification of biological solution • Definition of biological solution		• Principle abstraction • Reframing of solution		• Problem search • Problem definition		• Application of principle

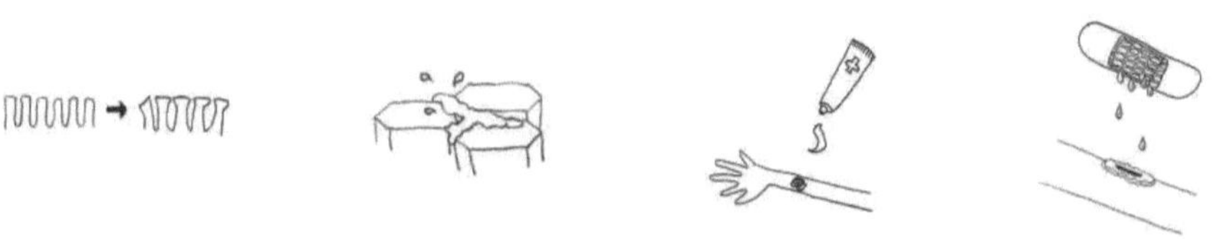

Description

This method is originally a description of observations from bio-inspired design projects. Nevertheless, it can be used as an orientation and guideline for biology-push projects. It can be divided into four phases: First, the biological system's solution is identified and defined. Second, the underlying principle is abstracted and reframed for the subsequent search. Third, the technical problem is searched and defined. Fourth, the abstracted principle is applied in a bio-inspired solution for the technical problem. Iterations are possible.

Advantages	Challenges
Practical details on the first phases of product development	Few details on the later phases of product development

Suitable for...	Biology push (☑)	Technology pull ☒

Creative ideation	● ● ○
Application of biological knowledge	● ● ●
Collaborative research projects	● ○ ○

Combines best with...	All other methods

Procedure for Collaborative Research Projects [VDI 2012]

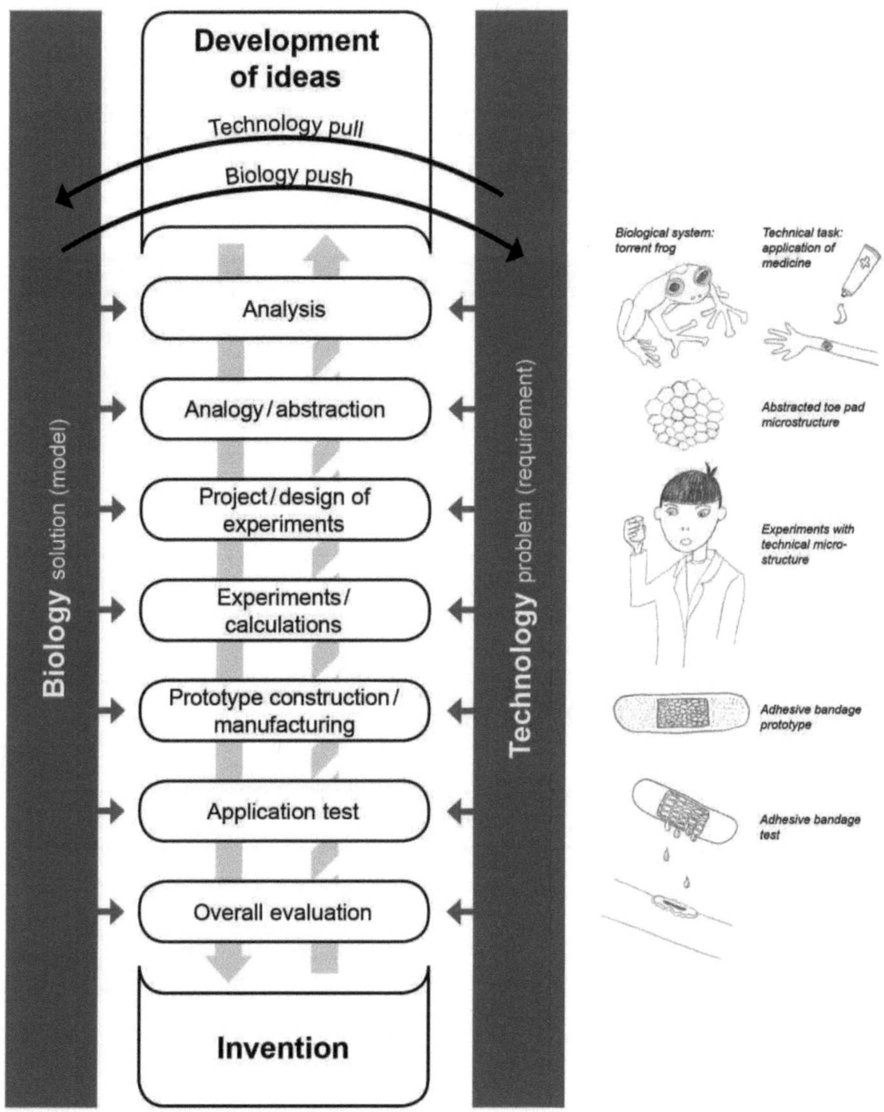

Description

The Procedure for Collaborative Research Projects proposed by the guideline VDI 6220 (VDI 2012) includes general steps of bio-inspired design, both for biology push and technology pull projects. The sequential visualization is an idealization of the process which, in reality, includes iterations. An ideation phase, which links a biological solution and a technical problem, is the starting point for the project. In the subsequent analysis phase, the biological system is analyzed using engineering and natural sciences methods to evaluate whether the biological solution can be transferred. An abstraction of the biological system and technical problem follows, in which transferable aspects are mapped. The subsequent steps of bio-inspired design are shown as an orientation for the further development of the analogy into an invention. However, these steps can differ from project to project. While the approach mainly describes technical product development, collaboration with biologists is beneficial and can alter the procedure.

Advantages	Challenges
Covers the whole product development process	Focus on the later phases of product development
	Mostly engineering viewpoint

Suitable for ...	Biology push ☑	Technology pull ☑

Creative ideation	○ ○ ○
Application of biological knowledge	○ ○ ○
Collaborative research projects	● ● ●

Combines best with ...	All other methods

PLANNING

TRIZ-based Procedure [Hill 1997]

Target Specification and Solution Development in Bio-Inspired Development Processes

1 TARGET SPECIFICATION

1.1 Market situation and customer demand

1.2 System analysis

1.3 Determination of state of the art

1.4 Generation analysis

1.5 Evaluation of status of evolution

1.6 Gathering of effectiveness factors and formulation of effectiveness function

1.7 Setting up requirements matrix and selection of relevant contradictions

1.8 Formulation of paradox demand

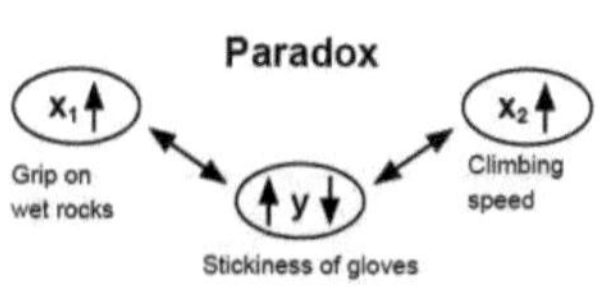

Ideal development direction

↓

Formulation of development task with inventive target

2 SOLUTION DEVELOPMENT

2.1 Determination of the basic function behind the paradox demand

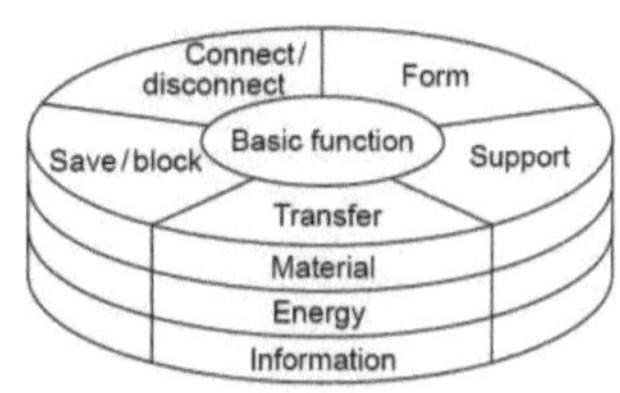

Orientation model of basic functions

2.2 Disclosure of relevant biological structures with the same or similar functional features

2.3 Composition of relevant structures in a chart and deduction of first solutions (principle solutions)

2.4 Development of these solutions into a complete technical solution according to the requirements (economic, technological, ecological, social, ...)

2.4.1 Variation and/or combination of relevant features

2.4.2 Evaluation of the elements of the solution or technical variations

2.5 Detailed design of the technical solution

↓

Technical solution

Description

The TRIZ strategy model combines methods of the TRIZ methodology with the search of biological principles in the catalog of basic functions [Hill 1997]. It covers the early phases of the product development process, keeping an engineering and functional viewpoint of the product. The TRIZ strategy model is divided into the main phases “target specification” and “solution development”. Through analysis of the market, the existing technical system, as well as its development status and requirements, a technical paradox may be formulated, which serves to define the goal. Then, solutions from the catalog are used, combined and evaluated to develop the solution.

Advantages	Challenges
Covers the early phases of the product development process	Experience and time effort needed to apply multi-step TRIZ methodology
	Mostly engineering viewpoint

Suitable for...	Biology push ☒	Technology pull ☑

Creative ideation	○ ○ ○
Application of biological knowledge	● ● ●
Collaborative research projects	● ○ ○

Combines best with...	Catalogs/Databases of Biological Principles for SEARCH

PLANNING

Procedure for Architecture and Industrial Design [VDI 2015]

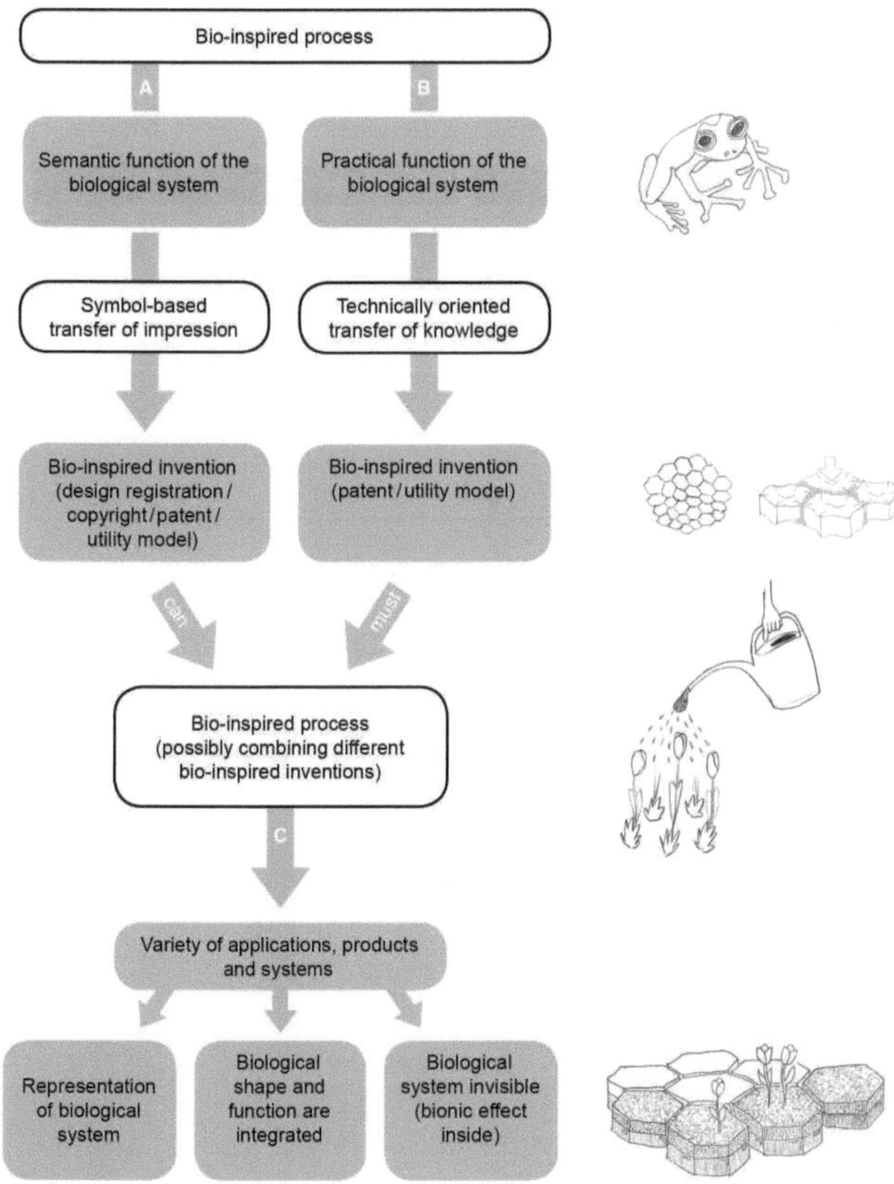

Description

The Procedure for Architecture and Industrial Design proposed by the guideline VDI 6226 (VDI 2015) focusses on architectural, civil engineering and industrial design projects. It therefore promotes an analysis of the interaction between semantic and practical functions. The model focusses on two major steps. In a first step, a "pool of (biological) ideas" is analyzed and abstract bio-inspired ideas are generated. To analyze the "pool of ideas" in a structured manner, comparisons can be conducted, using tables or morphological boxes. The resulting product, structure or component ideas are altered in a second step and transformed into the final application, product or system. In this step, several bio-inspired ideas can be combined. Moreover, the ideas can evolve in such a way that the bio-inspired aspects are invisible in the final result.

Advantages	Challenges
Illustrates the relationship between semantic and practical functions of bio-inspired design	
Useful for interdisciplinary collaboration	Few details on the procedure within the broad two steps

Suitable for ...	Biology push ☑	Technology pull ☑

Creative ideation	●●○
Application of biological knowledge	●●○
Collaborative research projects	●●○

Combines best with ...	All other methods

ABSTRACTION

Biocards [Lenau et al. 2011, 2015]

Phenomenon: Adhesion to wet and dry surfaces

Biological system: Organism: tree and torrent frogs

Photo/sketch:

Step-by-step abstraction

Description of the phenomenon and its underlying principle (biological terminology)

Biological phenomena:
Frog toe pads adhere and detach from dry and wet surfaces

Abstraction of the underlying principle – if possible in stages (technical terminology)

Biological mechanism:
The toe pads' micro- and nano structure play a role: The microstructure consists of hexamers. Channels between the hexamers store and release fluid under pressure. On nano-level the toes are built by pillar-shaped structures. Fluid is stored between the nano-pillars. Moreover, nano-pillars deform and thereby increase the contact area between frog toes and surface.

Depiction of the functional principle (in several stages)

Functional principle:
Several forces act in parallel: Channels between hexamers and nano-pillars absorb fluid causing a capillary effect. Pressure on micro and nano-pillar structure releases fluid increasing viscosity between the surfaces. The pressure furthermore increases the contact area causing higher friction and Van der Waals forces.

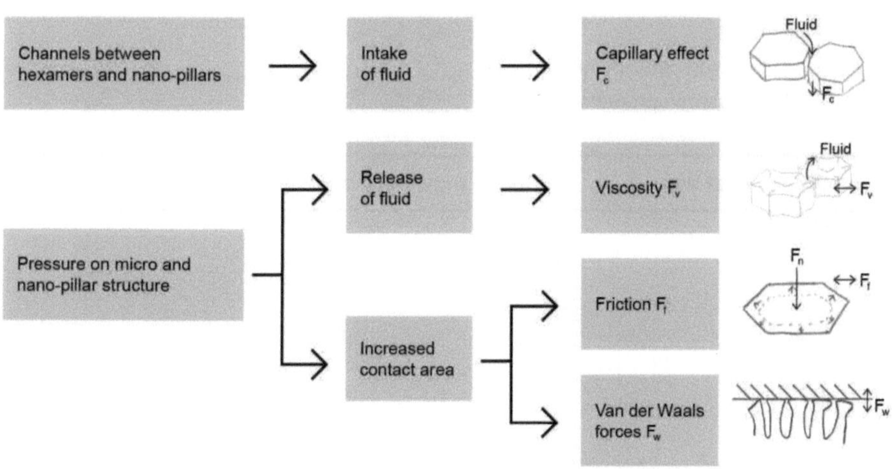

Description

Biocards are used to describe biological phenomena and to abstract them to a functional principle. While the biological phenomena are described using biological terminology, the functional principles are formulated in technical terminology. Thus, the preparation of a Biocard enables a step-by-step abstraction and shift from biological to technical viewpoint. To prepare Biocards, knowledge on the biological system and technical task is needed. This knowledge can be acquired from the results of a search (e.g. research publications, books etc.) or in a direct collaboration between engineers and biologists. Biocards are used to generate concrete technical solution ideas in different situations, e.g. by single designers or in interdisciplinary ideation workshops.

Advantages	Challenges
Simplification and visual representation of the principle	High degree of simplification: difficult to represent multi-facetted principles on one biocard
Step-by-step approach (concrete → abstract) facilitates the preparation of biocards	
Use of biological and technical terminology provides a "translation"	

Suitable for...	Biology push ☑	Technology pull ☑

Creative ideation	● ● ●
Application of biological knowledge	● ● ○
Collaborative research projects	● ● ○

Combines best with...	Technology Pull Procedure for PLANNING

ABSTRACTION

SBF Modeling [Goel et al. 2009, Yen et al. 2014, Georgia Institute of Technology 2016]

WHAT (structure): adhesive bandage

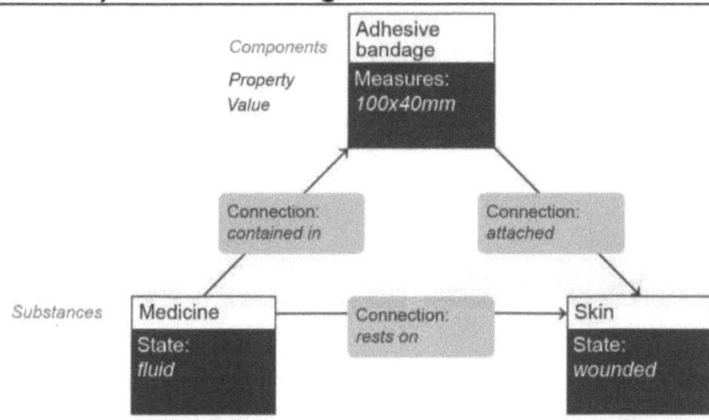

WHY (function): heal wound

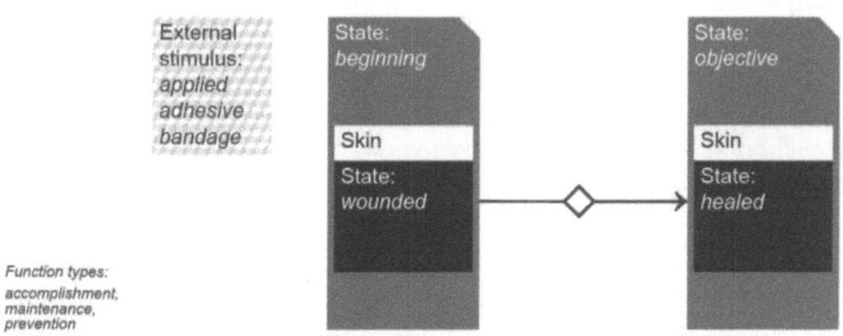

HOW (behavior): heal wound

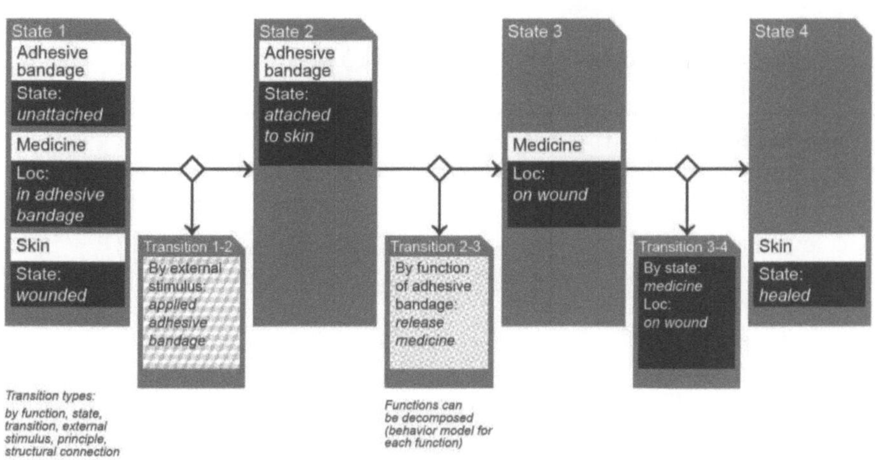

Description

SBF Modeling is based on concepts of *structure*, *behavior* and *function* which have been developed to represent technical solutions. SBF Modeling allows a detailed representation and explanation of systems from a functional viewpoint, i.e. they address the question "how does a system fulfils a certain purpose?". SBF Modeling starts with the concrete embodiment or structure of a system (*what*). Then the purpose or function of the system is represented by its beginning and objective state (*why*). Lastly, the behavior is modelled in detailed (*how*). It depicts how a system evolves from the beginning state to the objective state. It is possible to decompose the main function into sub-functions and to compose a behavior model for each function.

Advantages	Challenges
Detailed explanation of biological and technical systems' behavior as a chain of transitions	Time required, modeling experience and expert knowledge on the biological/technical systems
Step-by-step approach (concrete → abstract) which facilitates the preparation of SBF models	Choice of the adequate abstraction levels of functions [Yen et al. 2014]
Support of detailed understanding in interdisciplinary collaboration	Technical viewpoint and terminology

Suitable for...	Biology push ☑	Technology pull ☑

Creative ideation	○ ○ ○
Application of biological knowledge	● ● ●
Collaborative research projects	● ● ●

Combines best with...	All other methods

ABSTRACTION

SAPPhIRE [Chakrabarti et al. 2005, Ranjan et al. 2012]

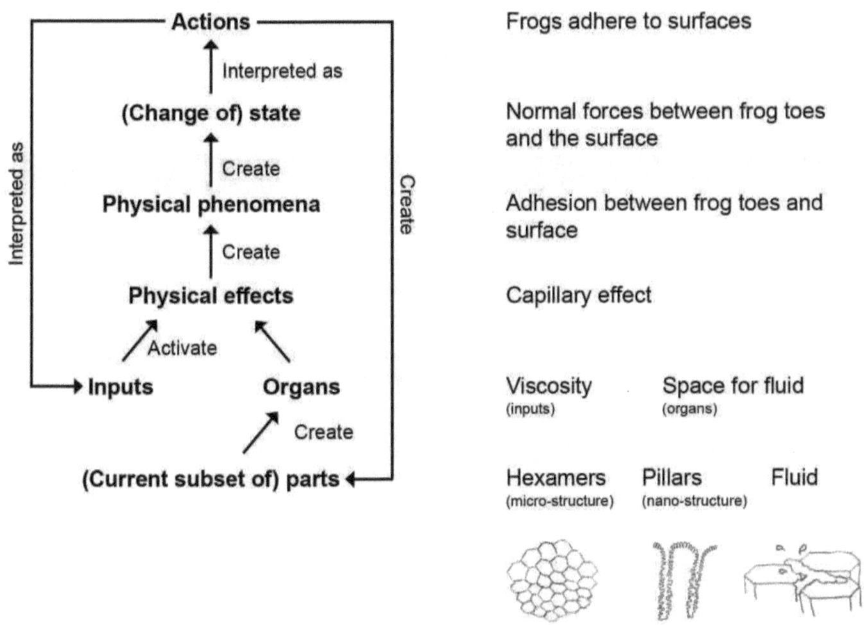

PARTS	Physical components and interfaces that constitute the object of interest and its surroundings.
ORGAN	A set of properties and conditions of the object and its surroundings required for an interaction between them.
INPUT	A physical variable that crosses the boundary of the object and is essential for an interaction between the object and its surroundings.
PHYSICAL EFFECT	A principle of nature that underlies and governs an interaction.
PHYSICAL PHENOMENON	An interaction between the object and its surroundings.
STATE	A property of the object (or its surroundings) that is involved in an interaction.
ACTION	An abstract description or high-level interpretation of an interaction between the object and its surroundings.

Description

SAPPhIRE models represent the causality of biological or technical systems resulting in the fulfillment of a purpose or function (action). SAPPhIRE is based on concepts of structure, behavior and function which have been developed to represent technical solutions. SAPPhIRE models show the causal chain between the concrete embodiment (parts) and the overall action. This chain includes properties (organs) of the system, as well as outside triggers (inputs). In combination, organs and inputs activate an abstract physical law (physical effect). This physical effect is responsible for a physical phenomenon causing a modification of the system's current state (change of state). This change of state implies the overall function of the system (action).

Advantages	Challenges
Detailed explanation of the function (action) of a biological/technical system	Requires time, modeling experience and expert knowledge on the biological/technical systems
Step-by-step approach (concrete → abstract) facilitates the preparation of SAPPhIRE models	Sequential representation of a system as a function (action).
Support of detailed understanding in interdisciplinary collaborations	Technical viewpoint and terminology

Suitable for...	Biology push ☑	Technology pull ☑

Creative ideation	○○○
Application of biological knowledge	●●●
Collaborative research projects	●●●

Combines best with...	All other methods

KoMBi [Hashemi Farzaneh 2015]

System description

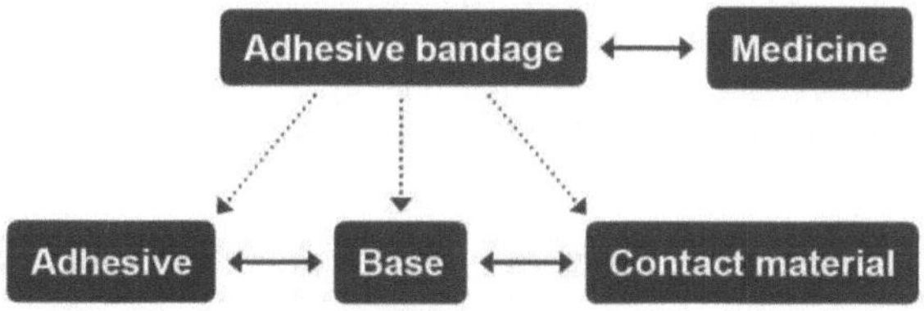

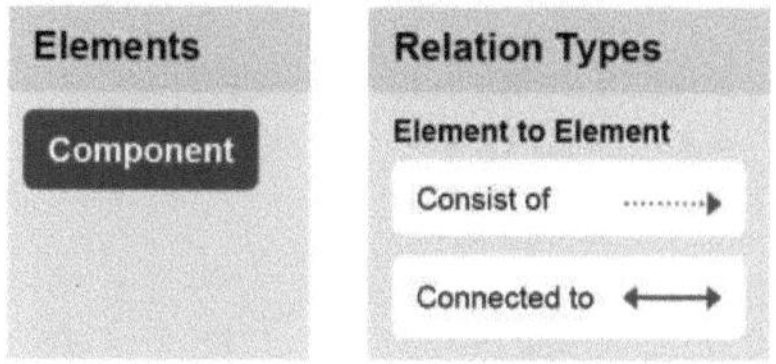

System behaviour & properties

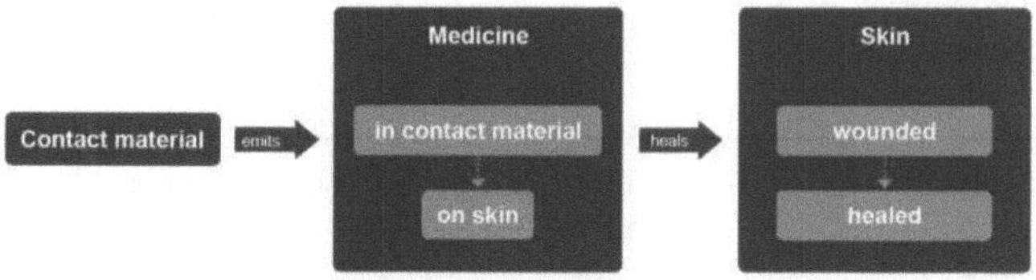

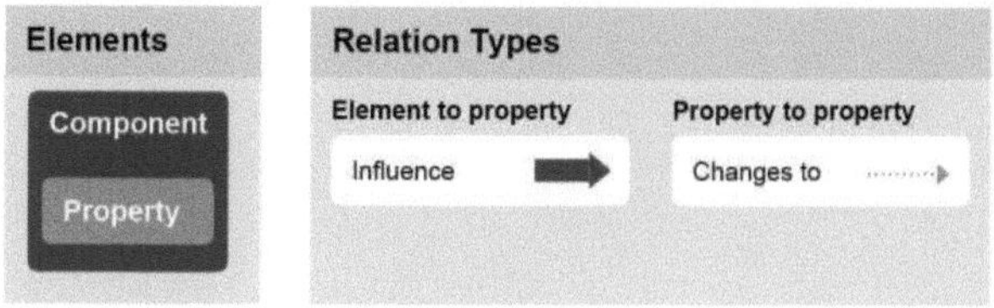

Description

KoMBi – a communication model for bio-inspired design – is based on features present in both biological and technical models. Both biological and technical systems can be represented by the same two-step modeling approach: In a first step, the system description is modeled. Five components of the system are named and connected by relations of the type "consists of" and "is connected to". In a second step, the system behavior and properties are represented. The user models those properties of the components that are relevant for the behavior. The changes are triggered by other components which "influence" the properties. Consequently, the affected properties "change to" different properties. The sequence of changes represents the system's behavior.

Advantages	Challenges
Explanation of the behavior of biological and technical systems as a chain of transformations	Requires time and effort to model the system
Incorporates both the biological and the technical viewpoint	The level of detail is lower than with comparable modeling approaches
Support of detailed understanding in interdisciplinary collaborations	

Suitable for ...	Biology push ☑	Technology pull ☑

Creative ideation	○ ○ ○
Application of biological knowledge	● ● ●
Collaborative research projects	● ● ●

Combines best with ...	Bio-Inspired Open Innovation: KoMBi Mapping for SEARCH

SEARCH

Research Publications [Kaiser et al. 2014]

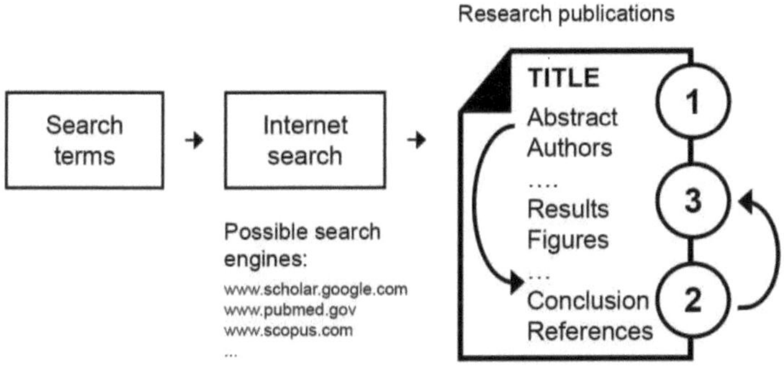

Description

Biological research publications serve as sources of information. To read them quickly, the titles and abstracts of the publications found are read first. Then, the conclusion provides the main results. The *results* and *figures* sections give a more detailed explanation. Authors' contact details are usually given. Moreover, references can provide further information on the biological or technical system.

Advantages	Challenges
Broad search space	Find adequate biological systems
Search for the latest biological and technical research	Handle the high number of results
	Understand and evaluate the suitability of the biological principle

Suitable for...	Biology push ☑	Technology pull ☑

Creative ideation ● ○ ○
Application of biological knowledge ● ● ●
Collaborative research projects ● ● ●

Combines best with...	BIOscrabble* Translation Technical/Biological Domain* Variation of Search Terms* * for SEARCH (**how** to search)

Technical Patents [Kaiser et al. 2014]

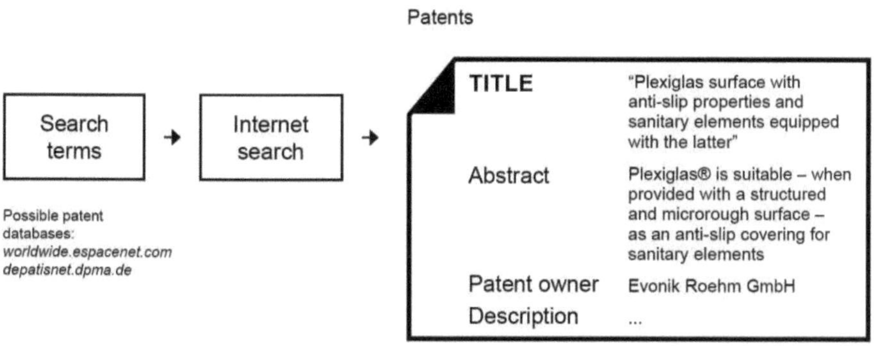

Description

Patents serve as source of information. To read them quickly, the titles and abstracts of the found publications are read first. Research on the owner of the patent can show application areas. The description shows the details of the patent.

Advantages	Challenges
Broad search space	Find adequate technical applications
	Handle the high number of results
	Understand and deduce a technical application

Suitable for...	Biology push ☑	Technology pull ☒

Creative ideation	● ○ ○
Application of biological knowledge	● ● ●
Collaborative research projects	● ● ●

Combines best with...	BIOscrabble for SEARCH (**how** to search)

SEARCH

Catalogs / Databases of Biological Principles

[Hill 1997, Gramann 2004, Löffler 2007]

Formulation of search terms → Search in catalog / database → Information on biological systems and principles

Catalog / database	Language	Search via…	N° of biological systems	Description of the biological principle	Picture
Catalog of basic functions (Hill1997)	German	Basic technical functions in combination with material / energy / information	~230	Description (functional characteristics)	Sketch explaining the principle
Association list (Gramann 2004)	German	Basic and concrete technical function + object / field / parameter	~300	☒	☒
Database of biomimetic effects (Löffler 2007)	German	"Principles" (all search terms are possible)	~180	Description of the biological principle, parameters, examples for technical application	Photo, sometimes sketches
"asknature. org (online database)"	English	Functions of the biomimicry taxonomy (all search terms are possible)	~1800	Description, bio-inspired products and application ideas, references	Photos

Description

Catalogs of biological principles provide descriptions of a number of biological systems. The technical task has to be formulated in abstract technical functions, with which the catalog can be searched. The biological systems in the catalogs are often well-understood and have already been explored for bio-inspired products.

Advantages	Challenges
Quick access to biological systems	No access to previously unexplored biological systems
No expert knowledge in biology required, the descriptions are formulated in technical terms	Technical task has to be formulated using the functions of the specific catalogue

Suitable for ...	Biology push ☒	Technology pull ☑

Creative ideation	●●●
Application of biological knowledge	●●○
Collaborative research projects	●○○

Combines best with ...	All other methods

Bio-inspired Open Innovation: KoMBi Mapping

[Hashemi Farzaneh 2015]

System behavior & properties

Biological system:
torrent frog's toe pads

Technical system:
adhesive bandage

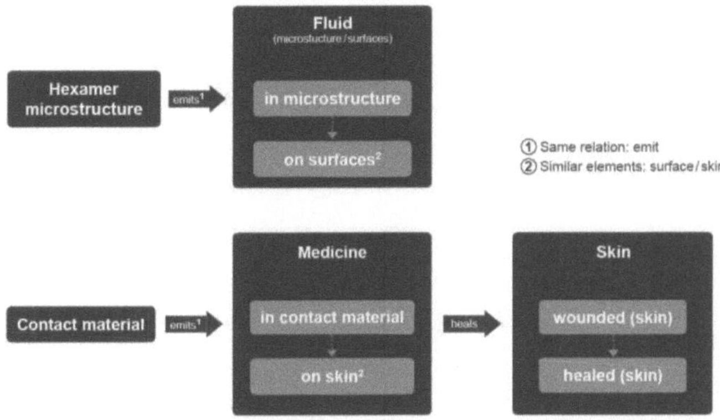

Description

To establish contact and the first steps of collaboration, engineers and biologists have to find suitable partners in the other discipline. KoMBi models can be used to map potential partners by identifying similar elements and relations. The visualization of the KoMBi model facilitates the understanding of the other discipline.

Advantages	Challenges
Automated matching of biological/ technical systems and potential collaboration partners	Time effort to model a biological or technical system with the KoMBi approach
Facilitates a first understanding between experts from the different disciplines	Simplified models can lead to misunderstandings

Suitable for ...	Biology push ☑	Technology pull ☑

Creative ideation	● ○ ○
Application of biological knowledge	● ● ●
Collaborative research projects	● ● ●

Combines best with ...	KoMBi for ABSTRACTION

SEARCH

Translation Technical / Biological Domain

[Nagel et al. 2010, Cheong et al. 2011, Baldussu 2014, p.86-87, 226-230]

Technical search term	Biological search term
Connect, couple, join, link	Biologically Meaningful Keywords: *extend, project, hold, stretch, overlap, activate, bind* Engineering-to-Biology Thesaurus: *recombination, mate, build, phosphorylate, bond, synthesis, latch, lock, extend, link, overlap, bind, adhere, bond, fuse, clamp, activate, project* Correlation Matrix NIST-BT: *attach permanently/temporarily, physically assemble structure*

Description

To translate a technical or biological search term to the other domain, several lists can be used. This supports the engineers or biologists to adapt their search query to the domain of search and search more successfully. All three translation approaches use the technical terms of the functional basis (Hirtz et al. 2002). For the List of Biologically Meaningful Keywords and the Engineering-to-Biology Thesaurus, biological terms were extracted from biology textbooks. The Correlation Matrix NIST-BT uses terms from the Biomimicry Taxonomy (The Biomimicry 3.8 Institute 2016). Therefore it is particularly useful for a search in the database asknature.org. All translation lists are included in the appendix.

Advantages	Challenges
Adapt search to the domain of interest	Select an adequate translation
Can be used in combination with other search methods	

Suitable for ...	Biology push ☑	Technology pull ☑

Creative ideation	● ● ○
Application of biological knowledge	● ● ●
Collaborative research projects	● ● ●

Combines best with ...	Catalogs / Databases of Biological Principles* Research Publication* Technical Patents* * for SEARCH (**where** to search)

Variation of Search Terms [Kaiser et al. 2014, Shu and Cheong 2014]

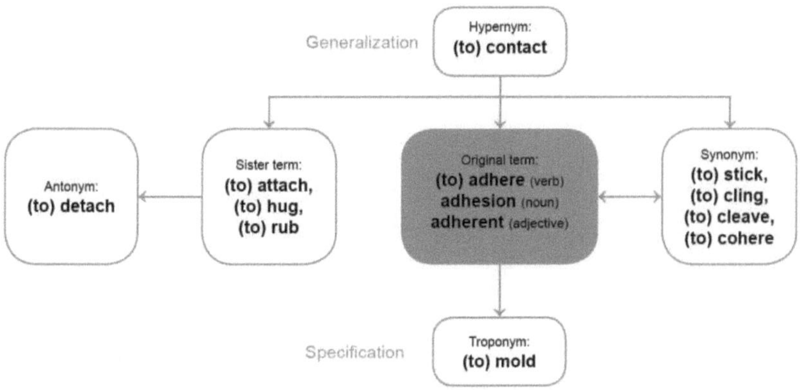

Description

To increase the number of search terms and results, the search terms can be varied: terms with similar meanings (synonyms), terms describing a certain manner of the original search term (troponyms), negations of the original search term and different parts of speech can be built. WordNet (http://wordnetweb.princeton.edu) and VerbNet (http://verbs.colorado.edu) provide lists of terms and their variations.

Advantages	Challenges
Broad search space	Handle the high number of results
Can be used in combination with other search methods	

Suitable for...	Biology push ☑	Technology pull ☑

Creative ideation	◉ ◉ ○
Application of biological knowledge	◉ ◉ ◉
Collaborative research projects	◉ ◉ ◉

Combines best with...	Research Publications* Technical Patents* * for SEARCH (**where** to search)

SEARCH

BIOscrabble [Helms 2016]

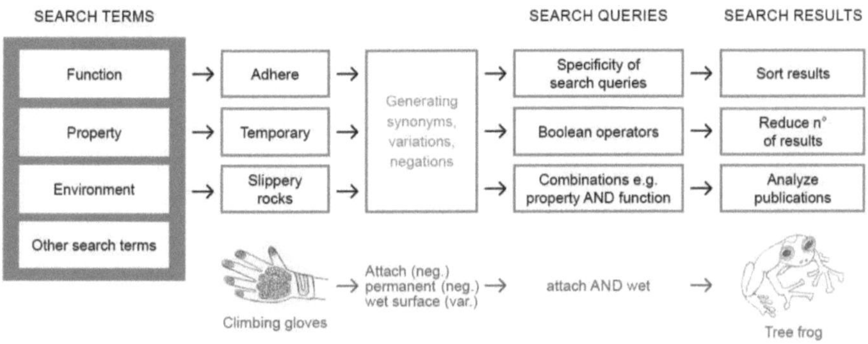

Description

A technical function is described using functions, desired properties and environmental influences. Synonyms, variations and negations are built. The search is conducted considering specifity of search queries and combining search terms via Boolean operators. Search results must be sorted, selected and analyzed.

Advantages	Challenges
Reflect on technical problem from different angles	
Broad search space	Handle the high number of results
Search for the latest biological research	Understand the biological research publications

Suitable for ...	Biology push (☑)*	Technology pull ☑

Creative ideation ○○○
Application of biological knowledge ○○○
Collaborative research projects ○○○

Combines best with ...	Research Publications* Technical Patents* * for SEARCH (**where** to search)

* Method can be adapted: a biological system is described instead of a technical problem, a patent database can be used instead of a publication database

ANALYSIS AND COMPARISON

Material Analysis [VDI 2013]

Integral Level
- Mechanical tests
- Analysis of the geometry and surface optimization

Macroscopic Level
- Light microscopy
- Mechanical tests
- Surface analysis

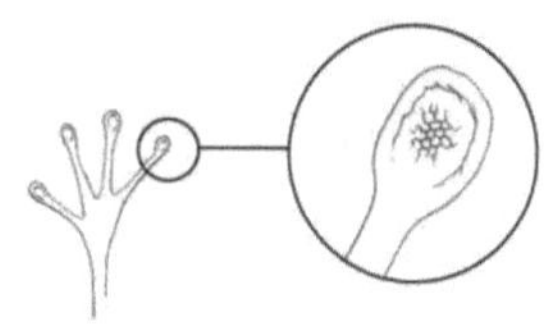

Microscopic Level
- Light microscopy
- Tomography
- Micromechanical tests

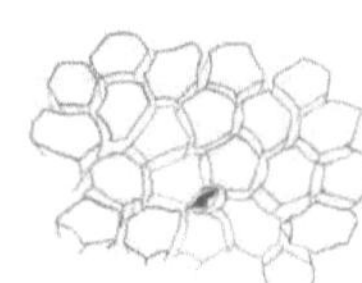

Submicroscopic Level
- Scanning electron microscopy (SEM ESEM)
- Transmission electron microscopy (TEM)
- Micromechanical tests
- Nanoindentation
- Scanning acoustic microscopy (SAM)
- Surface analysis (boundary surfaces, roughness, etc.)
- Multiphoton microscopy

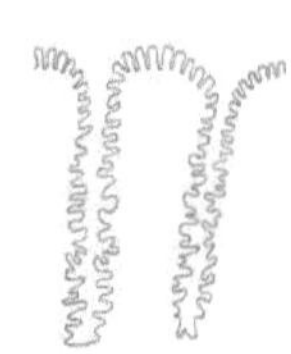

Biochemical Level
- Chemical analysis
- Spectroscopic analysis (FT-IR, Raman, UV)
- Atomic force microscopy (AFM)
- Transmission electron microscopy (TEM)
- X-ray scattering methods
- MMR

Description

As biological materials have a composite structure, they must be analyzed at different size scales. In addition, biological materials are often heterogeneous – an analysis must take into account irregularities of their structure and must be based on a number of samples. Moreover, the conditions of the biological system's environment must be considered, as the behavior of the biological material may change. Material tests must be conducted under the conditions of the envisioned technical solution. As such, parameters such as temperature, pressure and humidity must be considered.

Advantages	Challenges
Acquisition of a detailed understanding of structure and behavior of biological materials	Choice of adequate scales
	Consideration of the irregularities of biological materials

Suitable for...	Biology push ☑	Technology pull ☑

Creative ideation	○ ○ ○
Application of biological knowledge	○ ○ ○
Collaborative research projects	● ● ●

Combines best with ...	All other methods

Analysis of Scaling Effects

[Zerbst 1987; Baehr, Stephan 2011; Zierep, Bühler 2015]

Fluid flows	Reynolds number: $Re=\frac{d \cdot v \cdot l}{\eta}$ d: density, v: velocity, l: characteristic length, η: dynamic viscosity The Reynolds number has to be kept constant to ensure a similar fluid behavior. Moreover, the Reynolds number serves to differentiate between laminar (Re<1) and turbulent (Re>1) currents, which exhibit different behaviors.
Heat transfer	**Conduction:** • in solids/between solids: heat flow $\frac{Q}{t}=\frac{\lambda \cdot A \cdot \Delta T}{d}$ λ: conductivity of the material, A: area of heat transfer, ΔT: temperature difference, d: thickness of the material • between solids and fluids/between fluids (convection): The transfer of heat due to fluid flows depends on multiple variables such as dimensions and Reynolds number (for details see for example Baehr, Stephan 2011) **Radiation:** heat flow $\frac{Q}{t} = \varepsilon \cdot \sigma \cdot A \cdot T^4$ ε: emissivity (0...1) Stefan-Boltzmann constant $\sigma = 5.67 \cdot 10^{-8} \frac{W}{m^2 \cdot K^4}$ A: area, T: temperature
Forces	At a smaller scale, friction and intermolecular forces play an important role. At a larger scale, however, gravity gains in importance.

Scaling effects for transferring analogies from the torrent frog

Capillary effect:

$F_c \sim \frac{1}{r}$ (in this case r: width of the groove)

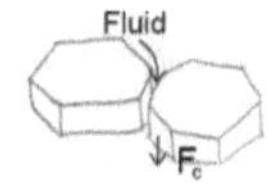

Friction: depends on friction coefficient. The friction coefficient is determined by the surface roughness and has a larger effect at smaller scales

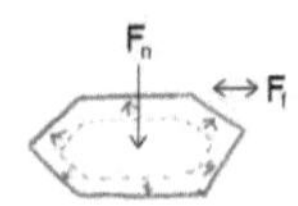

Van der Waals forces: act at the molecular level; their effect at the macro-level has to be measured or approximated

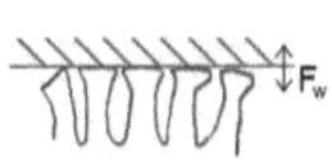

Description

To ensure the transferability of physical effects, the scales of the biological system have to be taken into account. The behavior of fluids, heat transfer and certain forces are nonlinear in relation to scales. In certain cases, e.g. to approximate the scaling effects regarding fluid flows, dimensionless variables such as the Reynolds number can be used. In any case, the complete system has to be analyzed carefully to ensure that all relevant scaling effects are taken into consideration. Often, simulations or testing with prototypes are necessary to determine scaling effects more adequately.

Advantages	Challenges
Ensuring the transferability of physical effects	Identification of all relevant factors which depend on the scales
Estimation of scaling effects using simplifications	Simulations and testing of prototypes can still be necessary

Suitable for...	Biology push ☑	Technology pull ☑

Creative ideation	●●○
Application of biological knowledge	●●●
Collaborative research projects	●●●

Combines best with...	Decision model similarity for ANALYSIS AND COMPARISON

ANALYSIS AND COMPARISON

Decision Model Similarity **[Zerbst 1987, Rechenberg 1978]**

Technical function F_t
Purpose to be achieved by a structure, form or principle in technology (e.g. absorbing loads, generating boost/ buoyancy, converting energy, measuring illuminance).

↓

Biological function F_b
Purpose to be achieved by a structure, form or principle in biology (e.g. carrying weight, flying, producing physical strength, perceiving light).

↓

Is F_t similar to F_b ? — no → **stop**

↓ yes

Function of torrent frog and adhesive bandage are similar: absorb/emit fluid

Technical constraints R_t
Invariable quantities to which the technical object is subjected (e.g. load types, flow velocities, material properties, wavelength spectrum).

↓

Biological constraints R_b
Invariable quantities given by the environment in which the living being exists (e.g. flow forces, wind velocities, material properties, solar spectrum).

↓

Is R_t similar to R_b ? — no → **stop**

↓ yes

Constraints due to fluid properties:
torrent frog: water, frog secretion
adhesive bandage: medicine, blood

Constraints due to material properties:
torrent frog: skin
adhesive bandage: synthetic material

Technical quality criteria Q_t
Which properties of the technical system influence its performance (e.g. weight, fuel consumption, flow resistance, instrument sensitivity)?

↓

Biological quality criteria Q_b
Which properties of the biological system influence its chance to survive (e.g. body weight, nutritional requirements, flow resistance, receptor sensitivity)?

↓

Is Q_t around Q_b ? — no → **stop**

↓ yes

Torrent frog and adhesive bandage:
Absorption/emission rates are influenced by:
- Pressure on the hexamer structure
- ...

DEVELOP BIO-INSPIRED SOLUTION

Description

The decision model supports the comparison between technical and biological systems. It supports a critical review of similarities and differences regarding functions, constraints and quality criteria. In a first step, the technical and biological functions or goals are compared. If they are similar, constraints that act upon the system are regarded. At this stage, a particular focus lies on dimensional effects: In many cases biological and technical system have different orders of magnitude. Designers have to pay special attention to ensure that the principle of the solution scales well. If the dimensions are similar as well, properties that influence the performance of both systems are used as quality criteria and compared.

Advantages	Challenges
Calls attention to the role of constraints and quality criteria, in particular if technical and biological system have different orders of magnitude	No observation of the abstraction level: more abstract analogies can be applied despite unsimilarity
Structures the procedure of critical review of similarities and differences between the technical and biological system	

Suitable for...	Biology push ☑	Technology pull ☑

Creative ideation	●○○
Application of biological knowledge	●●○
Collaborative research projects	●●●

Combines best with...	Dimensional Comparison for ANALYSIS AND COMPARISON

ANALYSIS AND COMPARISON

Four Box Method [Yen et al. 2014]

Environment	Function
Specification	Performance

Technical system: Climbing glove	Biological system: Torrent frog	Assessment
Environment: 10-25°C, 100% humidity	Environment: 15°C, 100% humidity	similar
Functions: Hold a portion (e.g. a quarter) of the climbing person's weight	Functions: Hold the frog's weight	not applicable
Specifications: Low force application for the gloves to adhere and detach	Specifications: Low force application to adhere and detach	same
Performance: Hold up to 20 kg → 1 N/cm^2	Performance: Hold up to 0.24 N/cm^2	different

Description

Technical and biological systems are described with regards to environment, functions, specifications and performance variables. This enables a side-by-side comparison of relevant variables and facilitates the evaluation of the biological inspiration's suitability for solving the technical task.

Advantages	Challenges
Structuring knowledge on biological and technical systems	Acquisition of information on the technical/biological systems
Side-by-side comparison highlights critical issues	

Suitable for ...	Biology push ☑	Technology pull ☑

Creative ideation	●●○
Application of biological knowledge	●●●
Collaborative research projects	●○○

Combines best with ...	All other methods

ANALYSIS AND COMPARISON

Expert-Novice-Layperson Dialog [Jordan 2008]

"Bridging the Knowledge and Terminology Gap"

Expert biology
layperson engineering

Novice biology/engineering
(e.g.: student, young professional, person with experience in bio-inspired design…)

Expert engineering
layperson biology

Description

Professional biologists and engineers are experts in their discipline but possibly lay-persons in the other discipline. The knowledge gap between them is therefore considerable. The inclusion of novices can bridge this knowledge gap. The novice can possibly explain knowledge better to the layperson than the expert.

Advantages	Challenges
Direct communication in this constellation can facilitate understanding considerably	Finding suitable novices: Ideally, they possess knowledge in both biology and technical product development
	Reservations of the experts regarding the novice → novice has to be well-prepared and plays an active role

Suitable for…	Biology push ☑	Technology pull ☑

Creative ideation	● ● ○
Application of biological knowledge	● ● ○
Collaborative research projects	● ● ●

Combines best with…	All other methods

TRANSFER

Analogy Model [VDI 2012]

Collecting aspects

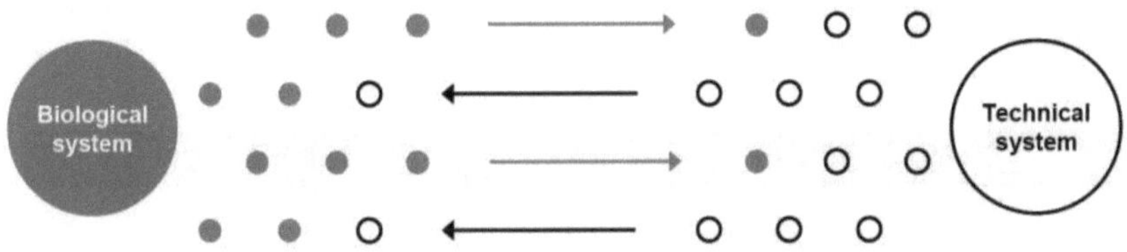

Building the relationship networks

Merging the relevant aspects

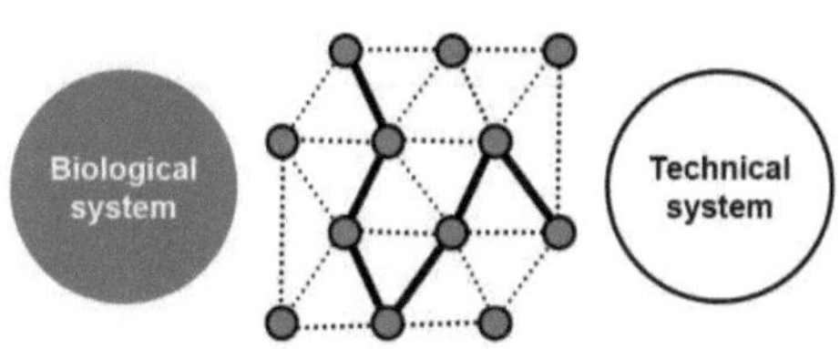

Bio-inspired system

Description

The analogy model defines an analogy as the transfer of similar relations between aspects of the biological and technical system. The first step of analogical transfer is therefore to collect relevant aspects of both systems. In a next step, the interactions and relations between the aspects are identified separately for the biological and the technical system. Then, the relationship networks are "superimposed". This comparison fosters the identification of different possible analogies.

Advantages	Challenges
Structured reflection on differences and similarities between the technical and the biological system	No support to collect the relevant aspects
Visualization of relevant aspects and their relations	

Suitable for...	Biology push ☑	Technology pull ☑

Creative ideation	● ● ○
Application of biological knowledge	● ● ○
Collaborative research projects	● ● ○

Combines best with...	All other methods

TRANSFER

Causal Relation Template [Cheong et al. 2010, Shu and Cheong 2014]

Biological system

Torrent frog's weight	*deforms*	*toe pads' micro- and nano-structure*	to	*enlarge*	*contact zone*
Subject	**Enabling function**	**Object B**		**Desired Function**	**Object A**

One-to-one mapping ↓

Bio-inspired technical system

Climber's weight	*deforms*	*glove's micro- and nano-structure*	to	*enlarge*	*contact zone*
Subject	**Enabling function**	**Object B**		**Desired Function**	**Object A**

Description

The causal relation template supports the formulation of a causal relation between two functions – enabling and desired function of a biological system. In a next step, the function has to be transferred to a bio-inspired solution.

Advantages	Challenges
Identification of transferable causal relations	Identification of more complex relations
	Fixation on transfering a single causal relation

Suitable for...	Biology push ☑	Technology pull ☑

Creative ideation	● ● ●
Application of biological knowledge	● ○ ○
Collaborative research projects	○ ○ ○

Combines best with...	All other methods

TRANSFER

Decision Model Abstraction Level

[Lindemann and Gramann 2004, p. 97-104]

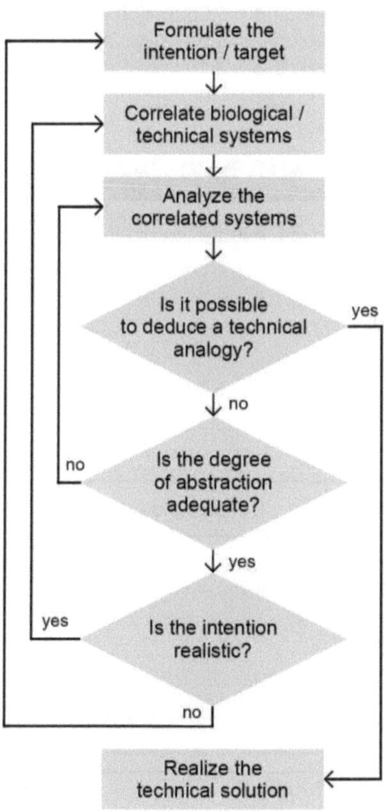

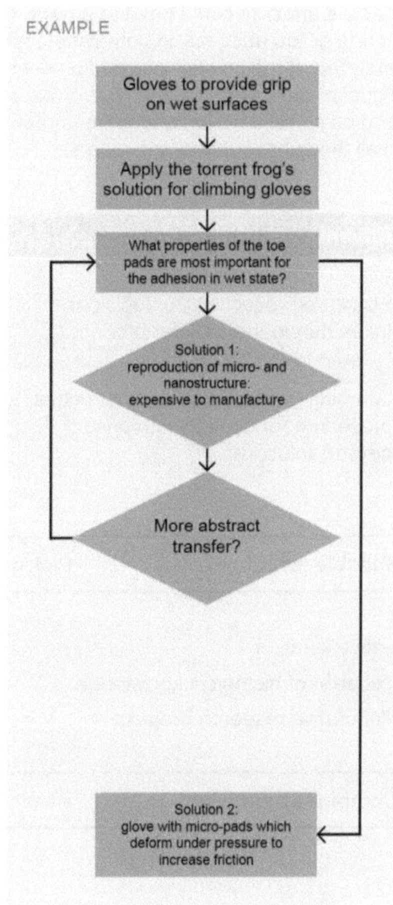

Description

The decision model supports the selection of an adequate abstraction level for transfer. It has been developed to be used in a technology pull approach, but can also be used for biology push approaches. Before the actual transfer, the goal of the transfer has to be formulated, biological and technical systems have to be correlated and analyzed. Then, if possible, an analogy is deduced. However, if this results in no feasible analogy, the decision model shows loops back to the previous steps. The reason for the lack of a feasible analogy could have to do with the abstraction level, which might be either too concrete or too abstract. In both cases, the biological and technical systems have to be reanalyzed. If the level of abstraction seems to be adequate, the next question concerns the goal of the transfer. If the goal is realistic, a different biological or technical system has to be correlated. In case of an unrealistic goal, however, this has to be reformulated and all the subsequent steps redone.

Advantages	Challenges
Structured reflection on decisions taken during the transfer process	No criteria to evaluate the bio-inspired analogies
Learning from failures – constructive procedure for dealing with unsuccessful solutions	

Suitable for...	Biology push ☑	Technology pull ☑

Creative ideation	● ● ○
Application of biological knowledge	● ● ○
Collaborative research projects	● ● ○

Combines best with...	All other methods

TRANSFER

Formulation of Different Analogy Types

[Mak and Shu 2004, Vakili et al. 2007, Sartori 2010]

Accuracy

Unrelated strategy no link to biological system

Use of a glue on the gloves

↓

Incorrect strategy misinterpretation of biological system

Coating of gloves with fine hair (similar to gecko)

↓

Incomplete strategy use of a "general principle" behind the strategy

Use of a rough material for gloves, e.g. leather

↓

Correct strategy use of "similar elements" compared to the biological system's strategy

Gloves with micro-/nano-structure similar to frog

Similarity

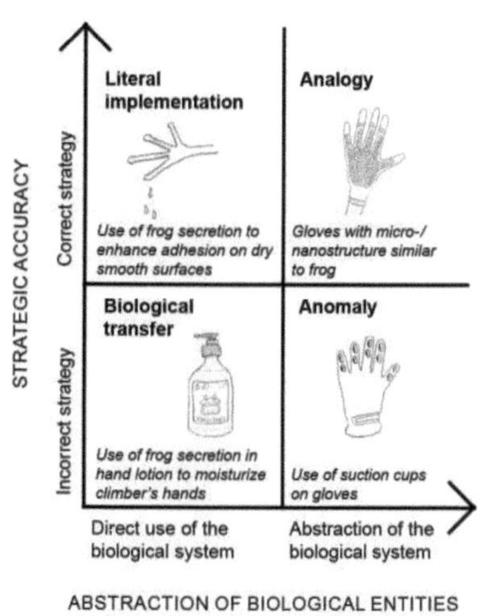

Elements of transfer

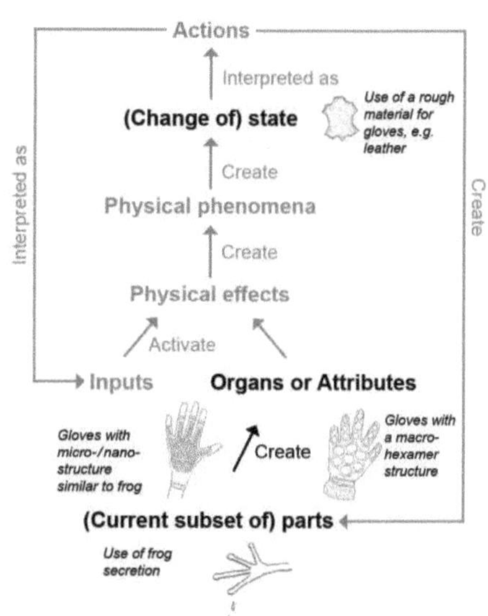

Description

Explicitly formulating different analogy types increases the number of bio-inspired solution ideas. The formulation of different analogy types pushes designers to consider different levels of abstraction and the degree of correctness (It has to be noted that incorrect transfers can also result in suitable solutions!). The formulation of different analogy types can therefore serve as a method of fostering creativity. Designers can generate a pool of solution ideas to choose the most suitable analogy and bio-inspired solution idea for further development.

	Challenges
Fosters reflection on the correctness and the degree of abstraction of analogies	Strictly generating only correct analogies can exclude potentially good solutions
Supports the transfer of several analogies	

Suitable for ...	Biology push ☑	Technology pull ☑

Creative ideation	● ● ●
Application of biological knowledge	● ● ○
Collaborative research projects	● ○ ○

Combines best with ...	SAPPhIRE for ABSTRACTION

TRANSFER

Biold [Hashemi Farzaneh, 2016]

Technical task

"Name the task of the desired technical solution."

Gloves with improved adhesion to wet surfaces.

"Which properties should it have?"

Hand size, weight to be supported etc.

Biological inspiration

CONCRETE LEVEL

1a) "Sketch the elements of the biological system which are relevant for the technical task and the desired properties."

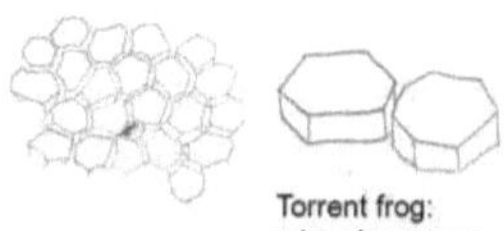

Torrent frog: micro-hexamers

1b) "How can these elements be transferred to a technical solution?"

Gloves with micro-hexamer structure

ABSTRACT LEVEL

2a) "Model the relations between the elements: How do they influence each other and change? What are their properties?"

2b) "How can the model of the biological system be transferred to a technical solution?"

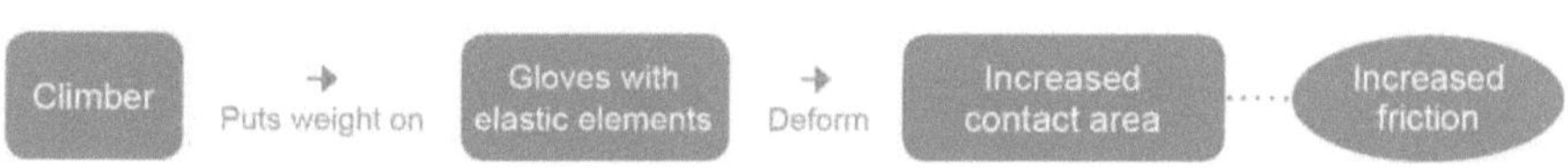

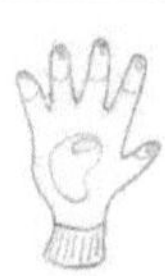

Description

Biold – a support for bio-inspired ideation – aims at workshops with engineers and biologists. It provides a template to represent the technical task, biological system and bio-inspired solution. To start with, the technical task has to be described. Then, the participants are asked to sketch a biological system, first on a concrete level. Based on this model, a technical analogy has to be derived and sketched. In a subsequent step, the biological system is modeled on a more abstract level using a graph. This second model serves to identify and transfer a second analogy, which can be either sketched or shown as a graph.

Advantages	Challenges
Fosters a structured, efficient approach in ideation workshops	Requires time and effort by the workshop participants
Supports the transfer of at least two analogies based on one biological system	Participants can lose motivation if they feel a lack of knowledge to represent the biological system
Communication support via visualizations in interdisciplinary collaborations	

Suitable for ...	Biology push ☒	Technology pull ☑

Creative ideation	●●●
Application of biological knowledge	●○○
Collaborative research projects	●○○

Combines best with ...	All other methods

TRANSFER

Synectics [Gordon 1961]

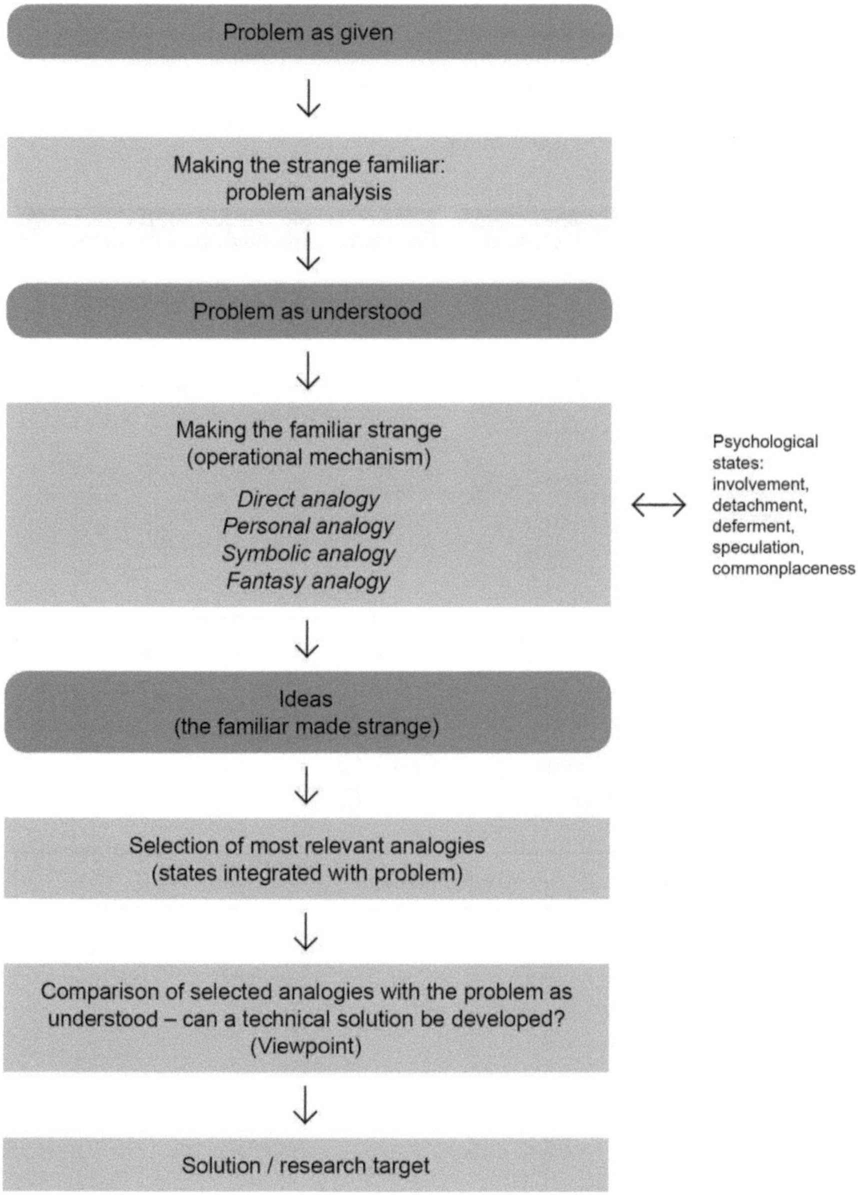

Description

The Synectics method is conducted in groups led by a moderator. For bio-inspired design, Synectics can be used to foster a collaboration between engineers and biologists. To start with, the group defines the problem, proposes spontaneous solutions and uses them to reformulate the problem. Then, the group generates different analogies: direct analogies, personal analogies, symbolic analogies and fantasy analogies. The most appropriate analogies are selected and compared to the technical problem. The result is a bio-inspired solution or a research target.

Advantages	Challenges
Supports the transfer of unobvious and distant analogies	Moderator with experience needed
Improvement of the problem definition	Multi-step method requires effort and motivation of the designers
	Relatively long duration (at least half a day)

Suitable for ...	Biology push ☒	Technology pull ☑

Creative ideation	● ● ●
Application of biological knowledge	● ○ ○
Collaborative research projects	● ○ ○

Combines best with ...	All other methods

8 References

Achleitner J (2014) Struktursimulation von geflochtenen Faserverbundbauteilen unter Berücksichtigung von Fertigungseinflüssen. Diploma thesis, Technical University of Munich

Adunka R (2016) Die Theorie des erfinderischen Problemlösens (TRIZ). In: Lindemann U (ed) Handbuch Produktentwicklung. Hanser, München, pp 673–689

Alberts B, Johnson A, Lewis J, Morgan D, Raff M, Roberts K, Walter P, Wilson J, Hunt T (2015) Molecular biology of the cell, Sixth edition. Garland Science Taylor and Francis Group, New York, NY

Altshuller GS (1984) Creativity as an exact science: the theory of the solution of inventive problems. Gordon and Breach, New York

Anderson JR (2009) Cognitive psychology and its implications, 7th edn. Worth Publishers, New York

Andreasen MM, Howard TJ, Bruun HPL (2014) Domain Theory, Its Models and Concepts. In: Chakrabarti A, Blessing LTM (eds) An Anthology of Theories and Models of Design. Springer London, London, pp 173–195

Ankeny RA, Leonelli S (2011) What's so special about model organisms? Studies in History and Philosophy of Science Part A 42:313–323. doi: 10.1016/j.shpsa.2010.11.039

Autumn K, Sitti M, Liang YA, Peattie AM, Hansen WR, Sponberg S, Kenny TW, Fearing R, Israelachvili JN, Full RJ (2002) Evidence for van der Waals adhesion in gecko setae. Proc Natl Acad Sci U S A 99:12252–12256. doi: 10.1073/pnas.192252799

Baehr HD, Stephan K (2011) Heat and Mass Transfer, 3., rev. ed. Springer, Berlin, Heidelberg

Bajaj C (2007) Geometric modeling and quantitative visualization of virus ultrastructure. In: Laubichler MD, Müller GB (eds) Modeling biology: Structures, behaviors, evolution. MIT Press, Cambridge, Mass., pp 115–137

Baldussu A (2012) Causal models for bio-inspired design: a comparison. In: Marjanovic D, Storga M, Pavkovic N, Bojcetic N (eds) Proceedings of DESIGN 2012, the 12th International Design Conference, Dubrovnik, Croatia. Faculty of Mechanical Engineering and Naval Architecture, University of Zagreb, Zagreb, Croatia, pp 717–726

Baldussu A (2014) A problem-solving methodology for the development of bio-inspired products. Doctoral thesis, Politecnico di Milano

Baluška F, Mancuso S (2016) Vision in Plants via Plant-Specific Ocelli? Trends Plant Sci 21:727–730. doi: 10.1016/j.tplants.2016.07.008

Barnard CJ, Gilbert FS, McGregor PK (2007) Asking questions in biology: A guide to hypothesis-testing, analysis and presentation in practical work and research, 3rd ed. Prentice Hall, Harlow

Bauer W, Dangelmeier M (2016) Virtuelles Engineering. In: Lindemann U (ed) Handbuch Produktentwicklung. Hanser, München, pp 939–952

H. Hashemi Farzaneh and U. Lindemann, *A Practical Guide to Bio-inspired Design*, https://doi.org/10.1007/978-3-662-57684-7

Beale A (2018) Best Climbing Gloves: The Top 5 of 2018. https://www.99boulders.com/best-climbing-gloves. Accessed 29 April 2018
Benyus JM (1997) Biomimicry: Innovation inspired by nature. Perennial, New York
Beyer C, Kochan D (2013) Innovationspotenziale der generativen Fertigung. In: Feldhusen J, Grote K-H (eds) Pahl/Beitz Konstruktionslehre: Methoden und Anwendung erfolgreicher Produktentwicklung, 8th edn. Springer, Berlin, pp 48–99
Biomimicry Institute (2016) Design spirals. http://ben.biomimicry.net/curricula-and-resources/university-curricula/reading-materials/. Accessed 4 July 2016
Blessing LT (1996) Comparison of design models proposed in prescriptive literature. In: Perrin J, Vinck D (eds) The role of design in the shaping of technology: COST A3/ COST A4 International Research Workshop. European Committee, Brussels, pp 187–212
BMW Group (2018) BMW Group plans Additive Manufacturing Campus: Technological expertise in industrial-scale 3D printing to be consolidated at new location, Munich
Böckh Pv, Wetzel T (2015) Wärmeübertragung: Grundlagen und Praxis, 6th edn. Springer Vieweg, Berlin, Heidelberg
Bogatyrev N, Bogatyreva O (2014) BioTRIZ: A Win-Win Methodology for Eco-innovation. In: Azevedo SG, Brandenburg M, Carvalho H, Cruz-Machado V (eds) Eco-Innovation and the Development of Business Models: Lessons from Experience and New Frontiers in Theory and Practice. Springer International Publishing, Cham, s.l., pp 297–314
Bono E de (2000) Six thinking hats, Rev. and updated ed. Penguin Books, London
Briggs JP (2002) The zebrafish: a new model organism for integrative physiology. Am J Physiol Regul Integr Comp Physiol 282:R3-9. doi: 10.1152/ajpregu.00589.2001
Brown T (2008) Design Thinking. Harvard Bussiness Review 86
Campbell NA, Reece JB, Urry LA, Cain ML, Wassermann SA, Minorsky PV, Jackson RB (2008) Biology, 8th edn. Pearson, San Francisco
Campbell NA, Urry LA, Cain ML, Wasserman SA, Minorsky PV, Reece JB (2018) Biology: A global approach, Eleventh edition, global edition. Pearson, New York, NY
Cardoso C, Badke-Schaub P (2009) Idea fixation in design: the influence of pictures and words. In: Grimheden M, Leifer L, Norell Bergendahl M (eds) Proceedings of the International Conference on Engineering Design (ICED09), Stanford. The Design Society, Glasgow, UK, pp 51–58
Chakrabarti A (2014) Supporting Analogical Transfer in Biologically Inspired Design. In: Goel AK, McAdams DA, Stone RB (eds) Biologically inspired design - computational methods and tools. Springer, London, pp 201–220
Chakrabarti A, Sarkar P, Leelavathamma B, Nataraju BS (2005) A functional representation for aiding biomimetic and artificial inspiration of new ideas. Arti-

ficial Intelligence for Engineering Design, Analysis and Manufacturing (AIEDAM) 19:113–132
Chan J, Fu K, Schunn C, Cagan J, Wood K, Kotovsky K (2011) On the benefits and pitfalls of analogies for innovative design: ideation performance based on analogical distance, commonness and modality of examples. Journal of Mechanical Design 133:81004
Cheong H, Chiu I, Shu LH (2010) Extraction and transfer of biological analogies for creative concept generation. In: ASME (ed) Proceedings of the ASME 2010 International Design Engineering Technical Conferences (IDETC) and Computers and Information in Engineering Conference (CIE), New York, p 29006
Cheong H, Chiu I, Shu LH, Stone RB, McAdams DA (2011) Biologically Meaningful Keywords for Functional Terms of the Functional Basis. Journal of Mechanical Design 133:21007. doi: 10.1115/1.4003249
Cheong, H., Shu, L. H., Stone RB, McAdams DA (2008) Translating terms of the functional basis into biologically meaningful keywords. In: ASME (ed) ASME 2008 International Design Engineering Technical Conferences (IDETC) and Computers and Information in Engineering Conference (CIE), New York, DETC2008-49363
Chi MT, Feltovich P, Glaser R (1981) Categorization and representation of physics problems by experts and novices. Cognitive Science 5:121–152
Christensen BT, Schunn CD (2007) The relationship of analogical distance to analogical function and preinventive structure: The case of engineering design. Memory & Cognition 35:29–38. doi: 10.3758/BF03195939
Chrysikou EG, Weisberg RW (2005) Following the wrong footsteps: fixation effects of pictorial examples in a design problem-solving task. J Exp Psychol Learn Mem Cogn 31:1134–1148. doi: 10.1037/0278-7393.31.5.1134
Crilly N (2015) Fixation and creativity in concept development: The attitudes and practices of expert designers. Design Studies 38:54–91. doi: 10.1016/j.destud.2015.01.002
Cross N (2001) Design Cognition: Results from Protocol and other Empirical Studies of Design Activity. In: Eastman CM, McCracken WM, Newstetter WC (eds) Design Knowing and Learning: Cognition in Design Education. Elsevier Science Ltd, New York, pp 79–103
Cross N (2008) Engineering design methods: strategies for product design, 4th edn. Wiley & Sons, Chichester
Dahl D, Moreau P (2002) The influence and value of analogical thinking during new product ideation. Journal of Marketing Research 39:47–60
Deldin J-M, Schuknecht M (2014) The asknature database: enabling solutions in biomimetic design. In: Goel AK, McAdams DA, Stone RB (eds) Biologically inspired design - computational methods and tools. Springer, London, pp 17–27
DIN 2330: 2013-07 (2013) Concepts and terms - General principles. DIN German Institute for Standardization, Berlin

Domke M-L, Hashemi Farzaneh H (2018) Research in bio-inspired design - what is its current focus? In: Proceedings of the International Conference on Design Creativity, Glasgow, UK

Dörner D (1987) Problemlösen als Informationsverarbeitung, 3rd edn. Kohlhammer, Stuttgart, Berlin, Cologne, Mainz

Drotlef D-M, Stepien L, Kappl M, Barnes WJP, Butt H-J, del Campo A (2013) Insights into the Adhesive Mechanisms of Tree Frogs using Artificial Mimics. Adv. Funct. Mater. 23:1137–1146. doi: 10.1002/adfm.201202024

Drotlef DM, Appel E, Peisker H, Dening K, del Campo A, Gorb SN, Barnes WJP (2015) Morphological studies of the toe pads of the rock frog, Staurois parvus (family: Ranidae) and their relevance to the development of new biomimetically inspired reversible adhesives. Interface Focus 5:20140036. doi: 10.1098/rsfs.2014.0036

Dunbar K (1999) How Scientists build models: invivo science as a window on the scientific mind. In: Magnani L, Nersessian N, Thagard P (eds) Model-based reasoning in scientific discovery. Plenum Press, New York, pp 89–98

Dunbar K, Blanchette I (2001) The in vivo/ in vitro approach to cognition: the case of analogy. TRENDS in Cognitive sciences 5:334–339

Dylla N (1990) Denk- und Handlungsabläufe beim Konstruieren. Doctoral thesis, Technical University of Munich

Ehrlenspiel K, Meerkamm H (2013) Integrierte Produktentwicklung: Denkabläufe, Methodeneinsatz, Zusammenarbeit, 5th edn. Hanser, München

Endlein T, Barnes WJP (2014) Wet Adhesion in Tree and Torrent Frogs. In: Bhushan B (ed) Encyclopedia of Nanotechnology. Springer Netherlands, Dordrecht, pp 1–20

Endlein T, Ji A, Samuel D, Yao N, Wang Z, Barnes WJP, Federle W, Kappl M, Dai Z (2013a) Sticking like sticky tape: tree frogs use friction forces to enhance attachment on overhanging surfaces. J R Soc Interface 10:20120838. doi: 10.1098/rsif.2012.0838

Endlein T, Barnes WJP, Samuel DS, Crawford NA, Biaw AB, Grafe U (2013b) Sticking under wet conditions: the remarkable attachment abilities of the torrent frog, Staurois guttatus. PLoS ONE 8:e73810. doi: 10.1371/journal.pone.0073810

Endlein T, Ji A, Yuan S, Hill I, Wang H, Barnes WJP, Dai Z, Sitti M (2017) The use of clamping grips and friction pads by tree frogs for climbing curved surfaces. Proc Biol Sci 284. doi: 10.1098/rspb.2016.2867

Faivovich J, Haddad CF, Garcia PC, Frost DR, Campbell JA, Wheeler WC (2005) Systematic review of the frog family Hylidae, with special reference to Hylinae: phylogenetic analysis and taxonomic revision. Bulletin of the American Museum of Natural History:1–240

Fellbaum C (ed) (1999) WordNet: An electronic lexical database, 2nd print. Language, speech, and communication. MIT Press, Cambridge, Mass

Fish FE, Beneski JT (2014) Evolution and Bio-Inspired Design: Natural Limitations. In: Goel AK, McAdams DA, Stone RB (eds) Biologically inspired design - computational methods and tools. Springer, London, pp 287–312

Fontoura Costa Ld (2007) The role of complex networks in biological modeling and systems biology. In: Laubichler MD, Müller GB (eds) Modeling biology: Structures, behaviors, evolution. MIT Press, Cambridge, Mass., pp 47–69

Forrester JW (1973) Industrial dynamics, Students' ed., 8. print. MIT Press, Cambridge, Mass.

Fu K, Chan J, Cagan J, Kotovsky K, Schunn CD, Wood K (2012) The meaning of "near" and "far": the impact of structuring design databases and the effect of distance of analogy on design output. In: Marjanovic D, Storga M, Pavkovic N, Bojcetic N (eds) Proceedings of DESIGN 2012, the 12th International Design Conference, Dubrovnik, Croatia, Vol 7: Human behaviour in design. Faculty of Mechanical Engineering and Naval Architecture, University of Zagreb, Zagreb, Croatia, pp 85–96

Furnham A (2000) The Brainstorming Myth. Business Strategy Review 11:21–28. doi: 10.1111/1467-8616.00154

Georgia Institute of Technology (2016) Learning about SBF models

Gero JS (1990) Design prototypes: a knowledge representation schema for design. AI magazine 11:26–36

Gianoli E, Carrasco-Urra F (2014) Leaf mimicry in a climbing plant protects against herbivory. Curr Biol 24:984–987. doi: 10.1016/j.cub.2014.03.010

Gick ML, Holyoak K (1983) Schema induction and analogical transfer. Cognitive Psychology 15:1–38

Gilovich T, Griffin D (2002) Introduction - Heuristics and biases: then and now. In: Gilovich T, Griffin D, Kahneman D (eds) Heuristics and biases - the psychology of intuitive judgement. Cambridge University Press, Cambridge, pp 1–18

Goel A, Rugaber S, Vattam s (2009) Structure, behavior, and function of complex systems: The structure, behavior, and function modeling language. Artificial Intelligence for Engineering Design, Analysis and Manufacturing 23:23–35

Goel AK (1997) Design, Analogy and Creativity. IEEE Intelligent Systems 12:62–70

Göker HM (1996) Einbinden von Erfahrung in das konstruktionsmethodische Vorgehen. Fortschritts-Berichte VDI Reihe 1, vol 268. VDI Verlag, Düsseldorf

Goldschmidt G, Sever AL (2011) Inspiring design ideas with texts. Design Studies 32:139–155. doi: 10.1016/j.destud.2010.09.006

Goldstein EB (2011) Cognitive psychology, 3. ed., international student ed., special ed. Wadsworth Cengage Learning, Belmont, Calif.

Gonçalves M (2016) Decoding designers' inspiration process. Doctoral thesis, Technical University of Delft

Gonçalves M, Cardoso C, Badke-Schaub P (2012) Find your inspiration: exploring different levels of abstraction in textual stimuli. In: Duffy A, Nagai Y,

Taura T (eds) Proceedings of The 2nd International Conference on Design Creativity (ICDC2012). The Design Society, Glasgow, UK

Gonçalves M, Cardoso C, Badke-Schaub P (2014) What inspires designers? Preferences on inspirational approaches during idea generation. Design Studies 35:29–53

Gonçalves M, Cardoso C, Badke-Schaub P (2016) Inspiration choices that matter: The selection of external stimuli during ideation. Des. Sci. 2:1. doi: 10.1017/dsj.2016.10

Gordon WJJ (1961) Synectics: The Development of Creative Capacity. Harper & Row, New York

Gramann J (2004) Problemmodelle und Bionik als Methode. Doctoral thesis, Munich, Technical University

Gürtler MR (2016) Situational Open Innovation: Enabling Boundary-Spanning Collaboration in Small and Medium-sized Enterprises. Doctoral thesis, Technical University of Munich

Haefner JW (2005) Modeling biological systems: principles and applications. Springer, New York

Hallihan GM, Cheong H, Shu LH (2012) Confirmation and Cognitive Bias in Design Cognition. In: Proceedings of the ASME 2012 International Design Engineering Technical Conferences (IDETC) and Computers and Information in Engineering Conference (CIE), New York, ASME Paper No. DETC2012-71258

Hashemi Farzaneh H (2016) Bio-inspired design: Ideation in collaboration between mechanical engineers and biologists. Doctoral thesis, Technical University of Munich

Hashemi Farzaneh H, Kaiser K, Lindemann U (2013) Influence of communication elements and cognitive effects on creative solution search in groups. In: Lindemann U, Srinivasan V, Kim YS, Lee SW, Ion B, Malmqvist J (eds) Proceedings of the 19th International Conference on Engineering Design (ICED13), Seoul. The Design Society, Glasgow, UK

Hashemi Farzaneh H, Helms K, Lindemann U (2015) Visual representations as a bridge for engineers and biologists in bio-inspired design collaborations. In: Weber C, Husung S, Cantamessa M, Cascini G, Marjanovic D, Srinivasan V (eds) Proceedings of the 20th International Conference on Engineering Design, Milan, vol 2. Design Society, Glasgow, UK, pp 215–224

Hashemi Farzaneh H, Kaiser MK, Lindemann U (2016) Selecting models from biology and technical product development for biomimetic transfer. In: Blessing L, Qureshi AJ, Gericke K (eds) The Future of Transdisciplinary Design. Springer, Heidelberg

Hatchuel A, Weil B (2003) A new approach of innovative design: an introduction to C-K theory. In: Folkeson A (ed) Research for practice - innovation in products, processes and organisations: ICED 03, 14th International Conference on Engineering Design ; 19 - 21 August 2003, The Royal Institute of Technology, Stockholm. Design Society, Glasgow

Hatchuel A, Le Masson P, Weil B (2011) Teaching innovative design reasoning: How concept–knowledge theory can help overcome fixation effects. AIEDAM 25:77–92. doi: 10.1017/S089006041000048X

Helms M, Vattam s, Goel AK (2009) Biologically inspired design: process and products. Design Studies 30:606–622

Helms MK (2016) Biologische Publikationen als Ideengeber für das Lösen technischer Probleme in der Bionik. Dissertation, Technical University of Munich

Helten K, Schenkl S, Lindemann U (2011) Biologizing product development - results from a student project. In: Chakrabarti A (ed) Proceedings of the International Conference on Research into Design. Research Publishing, Indian Institute of Science, Bangalore, India, pp 27–34

Herb R, Herb T, Kohnhauser V (2000) TRIZ: Der systematische Weg zur Innovation ; Werkzeuge, Praxisbeispiele, Schritt-für-Schritt-Anleitungen. mi Verl. Moderne Industrie, Landsberg/Lech

Hering E, Martin R, Stohrer M (2009) Taschenbuch der Mathematik und Physik, 5., aktualisierte und erw. Aufl. Springer Berlin Heidelberg, Berlin, Heidelberg

Hesse MB (1970) Models and analogies in science. University of Notre Dame Press, Notre Dame

Hill B (1997) Innovationsquelle Natur: Naturorientierte Innovationsstrategie für Entwickler, Konstrukteure und Designer. Shaker, Aachen

Hirtz J, Stone RB, McAdams DA, Szykman S, Wood KL (2002) A functional basis for engineering design: Reconciling and evolving previous efforts. Res Eng Design 13:65–82. doi: 10.1007/s00163-001-0008-3

Holyoak K, Koh K (1987) Surface and structural similarity in analogical transfer. Memory & Cognition 15:332–340

Hosoda N, Gorb SN (2012) Underwater locomotion in a terrestrial beetle: combination of surface de-wetting and capillary forces. Proc Biol Sci 279:4236–4242. doi: 10.1098/rspb.2012.1297

HSBC (2009) HSBC commercial: Washing machine used as lassi maker. https://www.youtube.com/watch?v=mZ8XBvLWL94. Accessed 13 April 2018

Hubka V, Eder WE (1988) Theory of technical systems: A total concept theory for engineering design, 2nd edn. Springer, Berlin

Iturri J, Xue L, Kappl M, García-Fernández L, Barnes WJP, Butt H-J, del Campo A (2015) Torrent Frog-Inspired Adhesives: Attachment to Flooded Surfaces. Adv. Funct. Mater. 25:1499–1505. doi: 10.1002/adfm.201403751

Jansson DG, Smith SM (1991) Design fixation. Design Studies 12:3–11

John G, Clements-Croome D, Jeronimidis G (2005) Sustainable building solutions: A review of lessons from the natural world. Building and Environment 40:319–328. doi: 10.1016/j.buildenv.2004.05.011

Jonkers HM (2007) Self Healing Concrete: A Biological Approach. In: Schmets AJM, van der Zwaag S (eds) Proceedings of the First International Conference on Self Healing Materials: 18 - 20 April, 2007, Noordwijk aan Zee, The Netherlands. Springer, Dordrecht, pp 195–204

Jordan A (2008) Methoden und Werkzeuge für den Wissenstransfer in der Bionik. Doctoral thesis, Otto-von-Guericke-Universität Magdeburg

Kaiser MK, Hashemi Farzaneh H, Lindemann U (2012) An Approach to Support Searching for Biomimetic Solutions Based on System Characteristics and its Environmental Interactions. In: Marjanovic D, Storga M, Pavkovic N, Bojcetic N (eds) Proceedings of DESIGN 2012, the 12th International Design Conference, Dubrovnik, Croatia. Faculty of Mechanical Engineering and Naval Architecture, University of Zagreb, Zagreb, Croatia, pp 969–978

Kaiser MK, Hashemi Farzaneh H, Lindemann U (2014) BIOscrabble - the role of different types of search terms when searching for biological inspiration in biological research articles. In: Marjanovic D, Storga M, Pavkovic N, Bojcetic N (eds) Proceedings of DESIGN 2014, the 13th International Design Conference, Dubrovnik, Croatia. Faculty of Mechanical Engineering and Naval Architecture, University of Zagreb, Zagreb, Croatia, pp 241–250

Kappl M, Kaveh F, Barnes WJP (2016) Nanoscale friction and adhesion of tree frog toe pads. Bioinspiration & Biomimetics 11:35003. doi: 10.1088/1748-3190/11/3/035003

Keshwani S, Chakrabarti A (2017) Influence of analogical domains and comprehensiveness in explanation of analogy on the novelty of designs. Res Eng Design 28:381–410. doi: 10.1007/s00163-016-0246-z

Keshwani S, Lenau TA, Kristensen SA, Chakrabarti A (2013) Benchmarking bio-inspired designs with brainstorming in terms of novelty of design outcomes. In: Lindemann U, Srinivasan V, Kim YS, Lee SW, Ion B, Malmqvist J (eds) Proceedings of the 19th International Conference on Engineering Design (ICED13), Seoul, Vol 7: Human behaviour in design. The Design Society, Glasgow, UK, pp 21–30

Keto JE (1960) Bionics - new frontiers of technology through fusion of the bio and physio disciplines (keynote address). In: Robinette JC (ed) Bionics Symposium: Living prototypes - the key to new technology, pp 7–12

Kipper K, Dang HT, Palmer M (2000) Class-Based Construction of a Verb Lexicon. In: Association for the Advancement of Artificial Intelligence (ed) AAAI-2000 Seventeenth National Conference on Artificial Intelligence

Kitano H (2002) Systems biology: a brief overview. Science 295:1662–1664. doi: 10.1126/science.1069492

Klemmstein W (2011) Bau und Lebensweise von Wirbeltieren. Unterrichtspraxis Biologie, vol 5. Aulis-Verl., Hallbergmoos

Knisely K (2005) A student handbook for writing in biology, 2. ed. Sinauer [u.a.], Sunderland Mass. u.a.

Koller R (2011) Konstruktionslehre für den Maschinenbau, 4th edn. Springer, Berlin

Kroll E, Jansson DG, Condoor SS (2001) Innovative conceptual design: Theory and application of parameter analysis. Cambridge University Press, Cambridge, New York

Lamprecht J (1999) Biologische Forschung: Von der Planung bis zur Publikation, Neubearb. Aufl. Studienhandbuch Biologie. Filander-Verl., Fürth

Lashin G (2013) Rechnerunterstützung in der Produktentwicklung und -konstruktion. In: Feldhusen J, Grote K-H (eds) Pahl/Beitz Konstruktionslehre: Methoden und Anwendung erfolgreicher Produktentwicklung, 8th edn. Springer, Berlin, pp 413–457

Laubichler MD, Müller GB (2007) Models in theoretical biology. In: Laubichler MD, Müller GB (eds) Modeling biology: Structures, behaviors, evolution. MIT Press, Cambridge, Mass., pp 3–9

Leedy PD (1989) Practical Research, 4th edn. Macmillan Publishing Company, New York

Lenau T, Dentel A, Ingvarsdóttir P, Gudlaugsson T (2010) Engineering design of an adaptive leg prosthesis using biological principles. In: Marjanovic D, Storga M, Pavkovic N, Bojcetic N (eds) Proceedings of DESIGN 2010, the 11th International Design Conference, Dubrovnik, Croatia. Faculty of Mechanical Engineering and Naval Architecture, University of Zagreb, Zagreb, Croatia, pp 331–340

Lenau T, Helten K, Hepperle C, Schenkl S, Lindemann U (2011) Reducing consequences of car collsion using inspiration from nature. In: Roozenburg N, Chen LL, Stappers PJ (eds) Proceedings of the 4th World Conference on Design Research (IASDR), Delft

Lenau T, Keshwani S, Chakrabarti A, Ahmed-Kristensen S (2015) Biocards and level of abstraction. In: Weber C, Husung S, Cantamessa M, Cascini G, Marjanovic D, Srinivasan V (eds) Proceedings of the 20th International Conference on Engineering Design, Milan, Vol 2: Design Theory and Research Methodology Design Processes. Design Society, Glasgow, UK, pp 177–186

Leonelli S (2008) Performing abstraction: two ways of modelling Arabidopsis thaliana. Biology and Philosophy 23:509–528

Lepora NF, Verschure P, Prescott TJ (2013) The state of the art in biomimetics. Bioinspiration & Biomimetics 8

Lieschke GJ, Currie PD (2007) Animal models of human disease: zebrafish swim into view. Nat Rev Genet 8:353–367. doi: 10.1038/nrg2091

Linde H, Hill B (1993) Erfolgreich erfinden: Widerspruchsorientierte Innovationsstrategie für Entwickler und Konstrukteure. Fachbuch Konstruktion. Hoppenstedt Technik Tab. Verl., Darmstadt

Lindemann U, Gramann J (2004) Engineering design using biological principles. In: Marjanovic D (ed) Proceedings of DESIGN 2004, the 8th International Design Conference, Dubrovnik, Croatia. Faculty of Mechanical Engineering and Naval Architecture, University of Zagreb, Zagreb, Croatia, pp 355–360

Lindemann U (2009) Methodische Entwicklung technischer Produkte, 3rd. Springer, Berlin

Lindemann U (2016) Kreativität in der Produktentwicklung. In: Lindemann U (ed) Handbuch Produktentwicklung. Hanser, München, pp 743–758

Löffler S (2009) Anwenden bionischer Konstruktionsprinzipe in der Produktentwicklung. Logos Verlag, Berlin

Lopez R, Linsey JS, Smith SM (2011) Characterizing the effect of domain-distance in design by analogy. In: ASME (ed) Proceedings of the ASME 2011 International Design Engineering Technical Conferences (IDETC) and Computers and Information in Engineering Conference (CIE). ASME, ASME Paper No. DETC2011-48428

Maas SA, Ellis BJ, Ateshian GA, Weiss JA (2012) FEBio: finite elements for biomechanics. J Biomech Eng 134:11005. doi: 10.1115/1.4005694

Mak TW, Shu LH (2004a) Abstraction of biological analogies for design. CIRP Annals - Manufacturing Technology 53:117–120

Mak TW, Shu LH (2004b) Use of biological phenomena in design by analogy. In: ASME (ed) Proceedings of the ASME 2004 International Design Engineering Technical Conferences (IDETC) and Computers and Information in Engineering Conference (CIE). ASME, New York, USA, ASME paper No. DETC2004/DETC-57303

McCarthy J (1956) The inversion of functions defined by Turing machines. Automata Studies:177–181

Meinke DW (1998) Arabidopsis thaliana: A Model Plant for Genome Analysis. Science 282:662–682. doi: 10.1126/science.282.5389.662

Meunier R (2012) Stages in the development of a model organism as a platform for mechanistic models in developmental biology: Zebrafish, 1970-2000. Stud Hist Philos Biol Biomed Sci 43:522–531. doi: 10.1016/j.shpsc.2011.11.013

Murphy SV, Atala A (2014) 3D bioprinting of tissues and organs. Nat Biotechnol 32:773–785. doi: 10.1038/nbt.2958

Nachtigall W (2002) Bionik: Grundlagen und Beispiele für Ingenieure und Naturwissenschaftler, 2. Auflage. Springer Berlin Heidelberg, Berlin, Heidelberg, s.l.

Nachtigall W (2010) Bionik als Wissenschaft. Springer, Heidelberg

Nachtigall W, Wisser A (2015) Bionics by examples: 250 scenarios from classical to modern times. Springer, Cham, Heidelberg, New York, Dordrecht, London

Nader W, Hill B (1999) Der Schatz im Tropenwald: Biodiversität als Inspirations- und Innovationsquelle. Shaker, Aachen

Nagel JKS (2014) A Thesaurus for Bioinspired Engineering Design. In: Goel AK, McAdams DA, Stone RB (eds) Biologically inspired design - computational methods and tools. Springer, London, pp 63–94

Nagel JKS, Stone RB, McAdams DA (2010) An Engineering-to-Biology Thesaurus for Engineering Design. In: ASME (ed) Proceedings of the ASME 2010 International Design Engineering Technical Conferences (IDETC) and Computers and Information in Engineering Conference (CIE), New York, ASME Paper No DETC2010/DTM-28233

National Center for Biotechnology Information (2018) PubMed. https://www.ncbi.nlm.nih.gov/pubmed/. Accessed 13 January 2018

Neinhuis C (2017) Innovations from the "ivory tower": Wilhelm Barthlott and the paradigm shift in surface science. Beilstein J Nanotechnol 8:394–402. doi: 10.3762/bjnano.8.41

Neugebauer E, Mutschler W, Claes L, Biberthaler P (2004) Von der Idee zur Publikation: Eine Anleitung zum erfolgreichen wissenschaftlichen Arbeiten ; 13 Tabellen. Thieme, Stuttgart

Nosonovsky M, Bhushan B (2012) Green Tribology: Biomimetics, Energy Conservation and Sustainability, 1. Aufl. Green Energy and Technology. Springer-Verlag, s.l.

Onstad DW, Meinke LJ (2010) Modeling Evolution of *Diabrotica virgifera virgifera* (Coleoptera Chrysomelidae) to Transgenic Corn With Two Insecticidal Traits. ec 103:849–860. doi: 10.1603/EC09199

Osborn AF (1963) Applied imagination: principles and procedures of creative problem-solving, 3rd edn. Charles Scribner's Sons, New York

O'Shea M, Halliday T (2001) Reptilien und Amphibien. Urania-Naturführer. Urania, Berlin

Pache M (2005) Sketching for Conceptual Design - Empirical Results and Future Tools. Doctoral thesis, Technical University of Munich

Pahl G, Beitz W (2013) Engineering design - a systematic approach, 2nd edn. Springer, London

Patankar NA (2004) Mimicking the lotus effect: influence of double roughness structures and slender pillars. Langmuir 20:8209–8213. doi: 10.1021/la048629t

Phoenix Bioinformatics Corporation (2017) The Arabidopsis Information Resource. http://www.arabidopsis.org. Accessed 10 December 2017

Ponn J, Lindemann U (2011) Konzeptentwicklung und Gestaltung technischer Produkte: Systematisch von Anforderungen zu Konzepten und Gestaltlösungen, 2nd edn. Springer, Heidelberg

Prechtl P, Burkard F-P (eds) (2015) Metzler Lexikon Philosophie: Begriffe und Definitionen, 3rd ed. J.B. Metzler'sche Verlagsbuchhandlung und Carl Ernst Poeschel Verlag GmbH, s.l.

Princeton University (2010) About WordNet. http://wordnet.princeton.edu/. Accessed 7 January 2018

Purves WK, Sadava D, Orians GH, Heller HC (2001) Life, The Science of Biology, 6th edn. Sinauer Associates, Sunderland

Purves WK, Sadava D, Orians GH, Heller HC (2004) Life: The Science of Biology, 7th edn. Sinauer Associates, Sunderland

Rechenberg I (1972) Evolutions-Strategie: Optimierung technischer Systeme nach Prinzipien der biologischen Evolution. Friedrick Frommann Verlag, Stuttgart

Rechenberg I (1978) Evolutions-Strategien. Medizinische Informatik und Statistik 8:84–114

Richter G (2003) Praktische Biochemie: Grundlagen und Techniken ; 193 Abbildungen, 19 Tabellen. Georg Thieme verlag, Stuttgart, New York

Rieg F, Nützel F, Wehmann C (2016) Numerische Simulationsverfahren. In: Lindemann U (ed) Handbuch Produktentwicklung. Hanser, München, pp 905–938

Roth K (2000) Konstruieren mit Konstruktionskatalogen - Bd. 2: Konstruktionskataloge, 3rd edn. Springer, Berlin

Rummel G (2014) SQAT® - ein strategisches Tool zur bionischen Innovation. In: Herstatt C, Kalogerakis K, Schulthess M (eds) Innovationen durch Wissenstransfer: Mit Analogien schneller und kreativer Lösungen entwickeln. Springer Fachmedien, Wiesbaden, pp 203–224

Sarkar P, Chakrabarti A (2008) The effect of representation of triggers on design outcomes. Artificial Intelligence for Engineering Design, Analysis and Manufacturing 22:101–116

Sartori J, Pal U, Chakrabarti A (2010) A methodology for supporting "transfer" in biomimetic design. Artificial Intelligence for Engineering Design, Analysis and Manufacturing 24:483–505

Schäfer M (2006) Computational Engineering - Introduction to Numerical Methods. Springer-Verlag Berlin Heidelberg, Berlin, Heidelberg

Schmutzler C, Teufelhardt S, Reinhart G, Zäh MF (2016) Neue Produktionstechnologien am Beispiel der additiven Verfahren. In: Lindemann U (ed) Handbuch Produktentwicklung. Hanser, München, pp 953–977

Scholz I, Barnes WJP, Smith JM, Baumgartner W (2009) Ultrastructure and physical properties of an adhesive surface, the toe pad epithelium of the tree frog, Litoria caerulea White. J Exp Biol 212:155–162. doi: 10.1242/jeb.019232

Serrat O (2017) The SCAMPER Technique. In: Serrat O (ed) Knowledge Solutions. Springer Singapore, Singapore, pp 311–314

Shu LH, Cheong H (2014) A natural language approach to biomimetic design. In: Goel AK, McAdams DA, Stone RB (eds) Biologically inspired design - computational methods and tools. Springer, London, pp 29–61

Silver PA (2011) Biology in This Century: An ancient science becomes the new technology. Harvard Magazine Sep-Oct 2011:72–73

Smith SM, Linsey J (2011) A Three-Pronged Approach for Overcoming Design Fixation. The Journal of Creative Behavior 45:83–91. doi: 10.1002/j.2162-6057.2011.tb01087.x

Srinivasan V, Chakrabarti A (2009) SAPPhIRE - an approach to analysis and synthesis. In: Grimheden M, Leifer L, Norell Bergendahl M (eds) Proceedings of the International Conference on Engineering Design (ICED09), Stanford. The Design Society, Glasgow, UK

Srinivasan V, Chakrabarti A, Lindemann U (2013) Understanding Internal Analogies in Engineering Design: Observations from a Protocol Study. In: Chakrabarti A, Prakash RV (eds) Proceedings of the International Conference on Research into Design (ICoRD'13), Indian Institute of Technology Madras, Chennai. Springer India, New Delhi, India, pp 211–222

Stachowiak H (1973) Allgemeine Modelltheorie. Springer, Wien

Streit B (2006) Biozahl 2006: 2-Millionen-Grenze erreicht. Natur und Museum 136:132–134

Suárez E (2007) Models and diagrams as thinking tools: the case of satellite DNA. History and Philosophy of the Life Sciences 29:177–192

Tarkan L (2003) Brain surgery without knife or blood, gains favor. New York Times

The Biomimicry 3.8 Institute (2016) Biomimicry taxonomy. http://www.asknature.org/article/view/biomimicry_taxonomy. Accessed 19 October 2017

The Ohio State University (2017) ABRC - Arabidopsis Biological Resource Center. https://abrc.osu.edu/. Accessed 10 December 2017

Thomas Publishing Company (2017) Online platform for supplier discovery and product sourcing. https://www.thomasnet.com/. Accessed 7 December 2017

Ulrich K, Eppinger S (2003) Product Design and Development, 3rd edn. McGrawHill, Singapore

Ulrich KT, Eppinger SD (2016) Product design and development, 6. ed. McGraw-Hill, New York, NY

University of Oregon (2017) The Zebrafish Information Network. https://zfin.org. Accessed 10 November 2017

University of Pennsylvania (2005) Verbnet: a class-based verb lexicon. http://verbs.colorado.edu/. Accessed 11 January 2018

Vakili V, Chiu L, Shu LH, McAdams DA, Stone RB (2007) Including functional models of biological phenomena as design stimuli. In: ASME (ed) ASME 2007 International Design Engineering Technical Conferences (IDETC) and Computers and Information in Engineering Conference (CIE), New York, ASME Paper No DETC2007-35776

Vandevenne D, Verhaegen P-A, Dewulf S, Duflou JR (2011) A scalable approach for the integration of large knowledge repositories in the biologically-inspired design process. In: Culley SJ, Hicks BJ, McAloone TC, Howard TJ, Dong A (eds) Proceedings of the 18th International Conference on Engineering Design (ICED11), Copenhagen. The Design Society, Glasgow

Vandevenne D, Verhaegen P-A, Dewulf S, Duflou JR (2015) A scalable approach for ideation in biologically inspired design. AIEDAM 29:19–31. doi: 10.1017/S0890060414000122

Vattam s, Wiltgen B, Helms M, Goel AK, Yen J (2011) DANE: Fostering Creativity in and through Biologically Inspired Design. In: Taura T, Nagai Y (eds) Design Creativity 2010. Springer, London, pp 115–122

Vattam SS, Helms ME, Goel A (2008) Compound analogical design: interaction between problem decomposition and analogical transfer in biologically inspired design. In: Gero JS, Goel AK (eds) Design Computing and Cognition '08. Springer, Dordrecht, pp 377–396

Vattam SS, Helms ME, Goel A (2009) Nature of creative analogies in biologically inspired innovative design. In: Grimheden M, Leifer L, Norell Bergendahl M

(eds) Proceedings of the International Conference on Engineering Design (ICED09), Stanford. The Design Society, Glasgow, UK, pp 255–264
Vattam SS, Helms ME, Goel A (2010) A content account of creative analogies in biologically inspired design. Artificial Intelligence for Engineering Design, Analysis and Manufacturing 24:467–481
VDI 2221 (1993) Methodik zum Entwickeln und Konstruieren technischer Systeme und Produkte. Association of German Engineers (VDI), Berlin
VDI 6220 (2012) Conception and Strategy - differences between biomimetic and conventional methods/ products. Association of German Engineers (VDI), Berlin
VDI 6223 (2013) Biomimetics - biomimetic materials, structures and components. Association of German Engineers (VDI), Berlin
VDI 6226 (2015) Biomimetics - architecture, civil engineering, industrial design: basic principles. Association of German Engineers (VDI), Berlin
Venter C, Cohen D (2004) The Century of Biology. New Perspectives Quarterly 21:73–77
Venter C, Cohen D (2014) The Century of Biology. New Perspectives Quarterly 31:28–37. doi: 10.1111/npqu.11423
Vincent JFV, Bogatyreva OA, Bogatyrev NR, Bowyer A, Pahl A-K (2006) Biomimetics: its practice and theory. Journal of the Royal Society Interface 3:471–482
Vogel S, Ferrari Ad (2013) Comparative biomechanics: Life's physical world, 2. ed. Princeton Univ. Press, Princeton, NJ
Vukusic P, Sambles JR (2003) Photonic structures in biology. Nature 424:852–855. doi: 10.1038/nature01941
Ware C (2012) Information Visualization - Perception for Design, 3rd edn. Morgan Kaufmann, Amsterdam
Weichert D (2005) Festigkeitslehre, 6th edn. Institute of General Mechanics, Aachen
Wilson JO, Rosen D, Nelson BA, Yen J (2010) The effects of biological examples in idea generation. Design Studies 31:169–186
Winkeljann B, Boettcher K, Balzer BN, Lieleg O (2017) Mucin Coatings Prevent Tissue Damage at the Cornea-Contact Lens Interface. Adv. Mater. Interfaces 7:1700186. doi: 10.1002/admi.201700186
Winkels R (2017) Schöne neue Wirklichkeit. https://www.helmholtz.de/technologie/schoene-neue-wirklichkeit/. Accessed 28 March 2018
Yen J, Helms M, Goel A, Tovey C, Weissburg M (2014) Adaptive evolution of teaching practices in biologically inspired design. In: Goel AK, McAdams DA, Stone RB (eds) Biologically inspired design - computational methods and tools. Springer, London, pp 153–199
Youmans RJ (2011) The effects of physical prototyping and group work on the reduction of design fixation. Design Studies 32:115–138

Youmans RJ, Arciszewski T (2014) Design fixation: Classifications and modern methods of prevention. AIEDAM 28:129–137. doi: 10.1017/S0890060414000043

Zerbst EW (1987) Bionik: Biologische Funktionsprinzipien und ihre technischen Anwendungen. Teubner-Studienbücher der Biologie. Vieweg +Teubner, Stuttgart

Zhao G, Perilla JR, Yufenyuy EL, Meng X, Chen B, Ning J, Ahn J, Gronenborn AM, Schulten K, Aiken C, Zhang P (2013) Mature HIV-1 capsid structure by cryo-electron microscopy and all-atom molecular dynamics. Nature 497:643–646. doi: 10.1038/nature12162

Zierep J, Bühler K (2015) Grundzüge der Strömungslehre: Grundlagen, Statik und Dynamik der Fluide : mit zahlreichen Übungen, 10., überarbeitete und erweiterte Auflage. Springer Vieweg, Wiesbaden

9 Appendix

H. Hashemi Farzaneh and U. Lindemann, *A Practical Guide to Bio-inspired Design*, https://doi.org/10.1007/978-3-662-57684-7

9.1 Collections of Biological Inspirations

Author/ Editor	Title	Year	Type of literature	Reference	Categories of bio-inspired design
Nachtigall, Wisser	Bionics by Examples	2015	Book	Cham: Springer	Historical examples, materials and structures, styling and design, construction and equipment, building and climatization, robotics and locomotion, sensors and neuronal control, anthropo- and biomedical technology, procedures and processes, evolution and optimization, systems and organization
Malik	Bionics-Fascination of Nature	2007	Book	Munich: MCB	316 image-text pages of examples: historical examples, robotics, locomotion, sensors, neuronal control, communication, organizational structures, self-organization, materials, structures, surfaces, evolution, optimization, intelligence and creativity
Kapsali	Biomimetics for Designers: Applying Nature's Processes & Materials in the Real World	2016	Book	London: Thames and Hudson	Historical examples, shapes, surfaces, structures, making, towards 4D designs
Xia	Biomimetic Principles and Design of Advanced Engineering Materials	2016	Book	New York: John Wiley & Sons	Material with mechanical and functional properties, surfaces and processes
Bennings	Bionics: Nature as a Model	1993	Book	Munich:Pro Futura	Image-text pages of examples in different disciplines: structures, surfaces, organizational structures, self-organisation, materials, energy supply, design, robotics, locomotion, light, sensors, manage compression, heat protection, evolution, optimization, processes, tools
Mano	Biomimetic Approaches for Biomaterials Development	2012	Edited book	Weinheim: Wiley-VCH	Material and surfaces

Author/ Editor	Title	Year	Type of literature	Reference	Categories of bio-inspired design
Bar-Cohen	Biomimetics: Biologically Inspired Technologies	2005	Edited book	Boca Raton: CRC Press,	Historical examples, material and surfaces, robotics and automation, evolution and optimization, strategies and mechanism, structures, sensors
Mukherjee	Biomimetics : Learning from Nature	2010	Edited book	London: InTech	Materials and surfaces, energy conversion, neuronal control, functions, chemical reactions, self-organization, robotics, locomotion, structures
Lakhtakia, Martín-Palma	Engineered Biomimicry	2013	Edited book	New York: Elsevier	Sensors, materials, robotics, fabrication, surfaces, structures, self-organization, evolution and optimization
Swiegers	Bioinspiration and Biomimicry in Chemistry: Reverse-Engineering Nature	2012	Edited book	New York: John Wiley & Sons	Self-assembled structures and systems, molecular machines, materials, chemistry, catalysis, surfaces, energy systems
Bar-Cohen, Breazeal	Biologically Inspired Intelligent Robots	2003	Edited book	Bellingham, WA: SPIE Press Monograph Vol. PM122, SPIE Publ.	Historical examples, robotics, materials, actuators, sensors, structures, functionality, control, intelligence, and autonomy
Allen	Bulletproof Feathers: How Science Uses Nature's Secrets to Design Cutting-Edge Technology	2010	Edited book	Chicago: University of Chicago Press	Marine dynamics, robotics, acoustics, cooperative behaviour, locomotion, materials and design

Author/ Editor	Title	Year	Type of literature	Reference	Categories of bio-inspired design
Gorb	Biologically-Inspired Systems	2011 2014 2015 2016 2017	Book Series	Cham: Springer	Description of different biological systems as inspiration for technical designs: Volume 1: Insect Biotechnology, Vilcinskas, 2011; Volume 2: Biotechnology of Silk, Asakura, Miller, 2014; Volume 3: Evolution of Lightweight Structures, Hamm, 2015; Volume 4: Biological Materials of Marine Origin, Ehrlich, 2015; Volume 5: Attachment Structures and Adhesive Secretions in Arachnids, Wolff, Gorb, 2016; Volume 6: Biomimetic Research for Architecture and Building Construction, Knippers, Nickel, Speck, 2016; Volume 7: Bio-inspired Structured Adhesives, Heepe, Xue, Gorb, 2017
Gerstman	Biological and Medical Physics, Biomedical Engineering	2012 2014 2015 2018	Book Series	Cham: Springer	Different topics of physical, chemical and biological sciences, for example: Biomedical Signals and Sensors I, Kaniusas, 2012; Bioinspiration, Liu, 2012 Mathematical Biophysics, Rubin et. al., 2014; Biomedical Signals and Sensors II, Kaniusas, 2015; Complex Fluids in Biological Systems, Spagnolie, 2015; Biomedical Signals and Sensors III, Kaniusas, 2018; Biomimetics, Bhushan, 2018
Kumar	Nanomaterials for the life science	2009 2010 2011	Book Series	Weinheim: Wiley-VCH	Volume 1: Metallic Nanomaterials, 2009; Volume 2: Nanostructured Oxides, 2009; Volume 3: Mixed Metal Nanomaterials, 2009; Volume 4: Magnetic Nanomaterials, 2009; Volume 5: Nanostructured Thin Films and Surfaces, 2010; Volume 6: Semiconductor Nanomaterials, 2010; Volume 7: Biomimetic and Bioinspired Nanomaterials, 2010; Volume 8: Nanocomposites, 2010; Volume 9: Carbon Nanomaterials, 2011; Volume 10: Polymeric Nanomaterials, 2011

Author/ Editor	Title	Year	Type of literature	Reference	Categories of bio-inspired design
Bar-Cohen	Biomimetics Series	2011 2016 2017 2018	Book Series	Boca Raton: CRC Press, Taylor & Francis Group	Volume 1: Biomimetics: Nature-based Innovation, Bar-Cohen, 2011; Volume 2: Ocean Innovation: Biomimetics Beneath the Waves, Anderson et al., 2016; Volume 3: Architecture Follows Nature-Biomimetic Principles for Innovative Design, Mazzoleni, 2017; Volume 4: Advances in Manufacturing and Processing of Materials and Structures, Bar-Cohen, 2018
Jiang, Guo, Liu	Biomimetic superoleophobic surfaces: focusing on their fabrication and applications	2014	Journal article	Journal of Materials Chemistry A, Vol.3(5), pp.1811-1827, doi: 10.1039/c4ta05582a	Superoleophobic surfaces
Vincent	Biomimetics-a review	2009	Journal article	Proceedings of the Institution of Mechanical Engineers, Part H: Journal of Engineering in Medicine, Vol.223(8), pp.919-939, doi: 10.1243/09544119JEIM561	Material, surface structures, mechanisms, hierarchy

Author/ Editor	Title	Year	Type of literature	Reference	Categories of bio-inspired design
Bhushan	Biomimetics: lessons from nature– an overview	2009	Journal article	Philosophical Transactions Of The Royal Society A-Mathematical Physical And, Vol.367(1893), pp.1445-1486, doi: 10.1098/rsta.2009.0011	Material, surfaces, hues, adhesion, motion, aerodynamics, energy productions, sensors, hierarchical organization
Bhullar, Lala, Ramkrishna	Smart Biomaterials – A review	2014	Journal article	Reviews On Advanced Materials Science, Vol.40(3), pp.303-314, http://www.ipme.ru/e-journals/RAMS/	Material and structures, smart biomaterials, self-healing, surfaces
Naik, Stone	Integrating biomimetics	2005	Journal article	Materials Today, Vol.8(9), pp.18-26, doi: 10.1016/S1369-7021(05)71077-4	Material and surfaces, structures

Author/ Editor	Title	Year	Type of literature	Reference	Categories of bio-inspired design
Yu, Lou, Fan, Zhang	Biomimetic optical materials: Integration of nature's design for manipulation of light	2013	Journal article	Elsevier Ltd., Progress in Materials Science, Volume 58, Issue 6, 825-873, doi: 10.1016/j.pmatsci.2013.03.003	Materials and surfaces, optical structures
Eadie, Ghosh	Biomimicry in textiles: past, present and potential. An overview	2011	Journal article	Journal of the Royal Society Interface, Vol.8(59), 761-775, doi: 10.1098/rsif.2010.0487	Material and functional surfaces, structures, colours, self-healing, thermal insulation in context to textiles
Ma	Biomimetic Materials for Tissue Engineering	2008	Journal article	Advanced Drug Delivery Reviews, Vol. 60, Issue 2, 184-198, doi: 10.1016/j.addr.2007.08.041	Nano-fibrous Materials, self-assembled structures, composite and nano-composite materials, surfaces, bioactive molecule delivery
Naleway, Taylor, Porter, Meyers, McKittrick	Structure and mechanical properties of selected protective systems in marine organisms	2016	Journal article	Materials Science and Engineering C,59, 1143 - 1167, doi: 10.1016/j.msec.2015.10.033	Structure and mechanical properties of marine organisms

Author/ Editor	Title	Year	Type of literature	Reference	Categories of bio-inspired design
Leong	Biomaterials	1980-present	Journal	ISSN: 0142-9612	Materials
Full	Bioinspiration & biomimetics : learning from nature	2006-present	Journal	ISSN: 1748-3190	Examples of different discplines
Anderson	Journal of Biomedical Materials Research Part A	1996-present	Journal	ISSN: 1552-4965	Materials
Gilbert	Journal of Biomedical Materials Research Part B: Applied Biomaterials	1996-present	Journal	ISSN: 1552-4981	Materials
Winnik	Langmuir	1985-present	Journal	ISSN: 0743-7463	Material and surfaces, structures, self-assembly, processes

Author/ Editor	Title	Year	Type of literature	Reference	Categories of bio-inspired design
Wagner	Acta Biomaterialia	2005-present	Journal	ISSN: 1742-7061	Material, surfaces and structures
Syngellakis	The International Journal of Design & Nature	2006-present	Journal	ISSN: 1755-7437	Design in nature, Nature and architectural design, Thermodynamics in nature, Evolutionary thermodynamics, Eco-informatics, Natural materials in engineering, Mechanics in nature, Dynamic modelling and ecosystems (etc.)
Boblan, Bannasch	First International Industrial Conference Bionik 2004	2004	Proceedings		Architecture, Design & Systematics; Biological Paradigms for Optimisation; Biomechatronics and Robotics; Fluid Dynamics; Functional Surfaces; Materials & Surfaces; Sensorics and Information Processing
	IEEE International Conference on Robotics and Biomimetics (ROBIO)	yearly	Conference		Robotics
	International Conference on Biomimetic and Biohybrid Systems, Living Machines	yearly	Conference		Living machines, robotics and locomotion, self-regulation, learning, memory, architecture, sensors
	International Conference on Intelligent Robotics and Applications (ICIRA)	yearly	Conference		Robotics, automation and intelligent systems

9.2 Catalogs/Databases of Biological Principles (Examples)

Functions (aggregated)	Sub group	Functions (of author)	Biological systems (examples)	Description	Source (database)
Move or stay put	Attach	Principles of joining	Suckers of octopus	Octopus possess different numbers and arrangement of suckers, which allow them to attach temporarily at an object	4)
		Attach permanently	Eggs of the asparagus beetle	Asparagus beetles attach their eggs permanently on plant surfaces	1)
	Move	Locomotion principles	Resilin of insects	Flying insects are able to move in/through gases due to the presence of resilin, an extremely elastic protein that stores the kinetic energy	4)
		Move (solids)	Organisms	Movement of organisms in/through liquids, gases or solids	3)
Protect from physical harm	Protect from living threats	Protection systems	Resolution of body contours	Some organisms disguise themselves by a color scheme matching the environment so that the body contours dis-solve to protect themselves from animals	4)
		Protect from plants/ microbes	White rock shell snail	The egg s urface of the white rock shell snail has special properties that protect the shell from microbial foul-ing	1)
	Protect from non-living threats	Protection systems	Lotus plants	Some plants, such as the lotus plant, have surfaces with wax layers that protect from dirt deposits	4)
	Manage structural forces	Shock absorption	Footpads of dog	Dogs have their own natural shock absorber - footpads - which manage structural forces like compression	4)
	Regulate physiol. processes	Regulate reproduction or growth	Cells of the freshwater common pond snail	Freshwater common pond snail regulates reproduction or growth by a clever mechanism of the cells to preserve vital resources	1)

Functions (aggregated)	Sub group	Functions (of author)	Biological systems (examples)	Description	Source (database)
Protect from physical harm	Prevent structural failure	Support/carry material	Insect trachea	Through spiral reinforcement the insect trachea prevents deformation with minimized material expenditure	2)
		Preserve	Plants	Plant resin prevents fracture/ rupture	3)
		Prevent buckling	Quills of porcupines	The thin walled quills of porcupines are built and optimized to prevent buckling	1)
Maintain community	Coordinate	Coordinate activities, coordinate systems, optimize shape/materials	Human skin	Human skin serves multiple functions such as to coordinate activities and systems	1)
	Cooperate	Cooperate within (eco)systems, co-operate/ compete between different species	Mutualism	In nature, there is a cooperation between different species of biological systems, for example some organism creating habitat for other species	1)
	Provide ecosystem services	Regulate climate, chemically break down inorganic compounds	Ants	Ants positively affect the properties of the soil and thereby reduce the carbon dioxide in the atmosphere which contributes to the regulation of the climate	1)

Functions (aggregated)	Sub group	Functions (of author)	Biological systems (examples)	Description	Source (database)
Modify	Modify physical state	Modify buoyancy	Swim bladders of fish	Swim bladders help fishes to swim with less energy in different depths by modifying their buoyancy	1)
		Deform	Insect wings	Insects modify size and shape of their wings after hatching from the cocoon	3)
		Form material	Birch leaf roller	Birch leaf rollers form a sinkhole out of birch leaves by modifying the physical state through cutting and rolling the leaves	2)
		Modify (colour)	Chameleons	Chameleons are able to change the colour of their skin rapidly	3)
	Modify chemical/ electrical state	Modify solubility (dissolving, emulsifying, precipitating, crystrallizing)	Lipid membranes in living cells	Lipid membranes in living cells have the ability to modify the solubility of fats and oils in water	1)
	Adapt/optimize	Adapt phenotype	Leaves of the Kukumakranka plant	The leaves of the Kukumakranka plant have the ability to adapt their phenotype to different climatic conditions	1)
	transform/ convert energy	Explosion mechanisms	Bombardier beetle	The bombardier beetle is able to transform thermal energy in its own combustion chamber by mixing and igniting a secretion mixture	4)
		Energy of a moving object	Muscles	Transform/ convert energy	3)

Functions (aggregated)	Sub group	Functions (of author)	Biological systems (examples)	Description	Source (database)
Make	Physically assemble	Bio-inspired architecture	Wasp's nest	The nest of wasps consists of a papery mass that helps the wasps to build a structure	4)
	Chemically assemble	Connect material	Spider silk	The silk of spiders to catch prey has a high surface tension which always keeps the silk tight	2)
	Chemically assemble	Produce	Synthesis	Chemically assembly with biochemical reactions (synthesis)	3)
Process information	Navigate	Measure/ recognize	Honeybees	Using electromagnetic waves honeybees are able to navigate through air	3)
	Send signals	Measure/ recognize	Fish	Some fishes are able to send electrical/ magnetic signals to communicate and navigate through water	3)
	Process signals	Measure/ recognize	Antennae of insects	Using their antennae, insects respond to signals	3)
	sense signals/ environmental cues	Transfer of information	Slit sensilla of spider	Spiders sense motion, sound and other vibrations with the slit sensilla that helps to stimulate the dendrites at low stresses caused by, for example, the movement of prey	2)
		Sensors	Compound eyes of insects	The temporal resolution and the wider field of view of compound eyes allow insects to sense light (visible spectrum) from the environment and thus avoid colliding with objects	4)

Functions (aggregated)	Sub group	Functions (of author)	Biological systems (examples)	Description	Source (database)
Break down	Chemically break down	Separate	Enzyme reaction	An enzyme reaction chemically breaks down polymers	3)
	Physically break down	Chopping devices	Radula of Patella limpets	The Patella limpet has one tooth, called radula, which has a special shape used for physically breaking down of living and non-living materials	4)
		Material separation	Chicken	Chicken physically break down non-living material when they use their tooth called "egg tooth" to open the eggshell	2)
Get, store, or distribute resources	Capture, absorb or filter	Distribution systems	Pigments in the oriental hornet´s cuticle	Hornets have special photovoltaic pigments in the cuticle that help them to absorb solar energy and turn it into electrical energy	4)
		Transfer of material	Tongue of woodpecker	The woodpeckers capture solids and liquids with their tongue	2)
	Store	Collect	Pelt/ feathers	The pelt and feathers of animals have the function to store thermal energy	3)
	Distribute	Distribute	Flying seeds	Some plants have seeds that are able to fly to distribute those solids (seeds) in the surrounding	3)
	Expel	Expel solids	Glands of marine birds	Marine birds are able to drink saline water of the sea because they expel the salt via glands	1)

9.3 Translation Technical/Biological Domain

Technical terms according to the Functional Basis (Hirtz et al. 2002), biological terms according to the Engineering-to-Biology-Thesaurus (Nagel et al. 2010), Biologically Meaningful Keywords (Cheong et al. 2011), and the Correlation Matrix NIST-BT (Baldussu 2014, p. 86-87, p. 226-230)

Technical Terms			Biological Terms			
Functional Basis			Eng.-to-Bio. Thesaurus	Biologically Meaningful Keywords	Correlation Matrix NIST-BT	
Class	*Secondary*	*Tertiary*	*Function Correspondents*	*Keywords (% of colloc., # of matches)*	*Group*	*Sub-Group*
Branch	Separate		Bleaching, meiosis, abscission, mitosis, segment, electrophoresis, dialysis, denature, free, detach, release	Speciate (%68,#66), Diverge (%44, #39), Segregate (%35, #34), Furrow (%33, #9), Evolve (%18, #424), Denature (%17, #36), Grow (%16, #796), Reproduce (%14, #537), Cleave (%14, #80), Surround (%11, #209), Stimulate (%9, #289), Contract (%3, #226), Activate (%2, #256), Retract (%14, #7), Bend (%12, #33), Fold (%8, #74)		
		Divide	Division, prophase, metaphase, anaphase, cleave, cytokinesis		Break Down	Chemically break down, Physically break down
		Extract		Collect (%11, #72), Extract (%10, #61), Trap (%8, #49), Delete (%7, #43), Degrade (%6, #36), Beat (%5, #39), Separate (%3, #308)	Get, store, or distribute resources	Capture, absorb, or filter
		Remove	Deoxygenate, filtrate, liberate, expulsion, evacuate		Get, store, or distribute resources	Capture, absorb, or filter
	Distribute		Circulate, diffuse, exchange, disperse, scatter, spread, spray	Hydrolyze (%41, #75), Burst (%32, #31), Discharge (%29, #14), Stimulate (%26, #289), Circulate (%26, #164), Fuse (%23, # 120), Secrete (%21, #232), Concentrate (%21, #58), Pass through (%20, #139), Break down (%20, #125), Diffuse (%15, #238), Stretch (%15, #89), Bind (%14, #483), Segregate (%12, #34), Change shape (%9, #71), Lyse (%26, #23), Decompose (%13, #31), Condensate (%6, #16), Fold (%5, #74)	Get, store, or distribute resources	Distribute

Technical Terms			Biological Terms			
Functional Basis			Eng.-to-Bio. Thesaurus	Biologically Meaningful Keywords	Correlation Matrix NIST-BT	
Class	*Secondary*	*Tertiary*	*Function Correspondents*	*Keywords (% of colloc., # of matches)*	*Group*	*Sub-Group*
Channel	Import		Absorb, attract, consume, inhale, intake	Osmose (%16, #31), Pass through (%15, #139), Squeeze (%14, #21), Diffuse (%7, #238), Insert (%5, #132), Release (%4, #508), Secrete (%3, #232), Transport (%3, #283), Fold (%1, #74)		
	Export		Bind, block, breakdown, excrete, inactivate, repel	Contract (%1, #266), Inactivate (%6, #52), Denature (%6, #36), Attach (%3, #200), Break down (%2, #125), Bind (%1, #483), Cleave (%1, #80), Excrete (%1, #111), Fuse (%1, #120)	Get, store, or distribute resources	Expel
	Transfer		Migrate, transfer	Conjugate (%60, #32), Beat (%41, #39), Transport (%27, #283), Couple (%22, #58), Break (%8, #196), Pollinate (%7, #74), Bind (%6, #483), Attract (%3, #96), Change shape (%52, #71), Organize (%10, #134) Shift (%7, #67)	Process information	Navigate
		Transport	Circulate, conduct, diffuse, pump, shift, displace, fly, swim, jump, bounce	Transport (%19, #283), Transduce (%10, #99), Communicate (%6, #109), Bind (%6, #483), Extend (%3, #95), Collect (%3, #72), Stimulate (%2, #289), Contract (%1, #226), Pollinate (%9, #74), Disperse (%4, #123)	Move or Stay Put	Move
		Transmit	Communicate, transduce	Contract (%12, #226), Transduce (%8, #99), Communicate (%6, #109), Conduct (%1, #106)		

Technical Terms			Biological Terms			
Functional Basis			Eng.-to-Bio. Thesaurus	Biologically Meaningful Keywords	Correlation Matrix NIST-BT	
Class	*Secondary*	*Tertiary*	*Function Correspondents*	*Keywords (% of colloc., # of matches)*	*Group*	*Sub-Group*
Channel	Guide		Orient, position, slide, tunnel	Communicate (%6, #109), Extend (%3, #95), Transport (%3, #283), Arrange (%3, #112), Hold (%2, #124)		
		Translate	Synthesize, transcribe	Transcribe (%27, #347), Synthesize (%12, #310)		
		Rotate	Oscillate, spin, turn, swivel, roll	Wind (%4, #26)		
		Allow DOF	Articulate			
Connect	Couple		Recombination, mate, build, phosphorylate, bond, synthesis, latch, lock, extend, link, overlap	Extend (%15, #95), Project (%14, #50), Hold (%14, #124), Stretch (%13, #89), Overlap (%10, #29), Activate (%7, #256), Bind (%4, #483)		
		Join	Bind, adhere, bond, fuse		Make	Physically assemble
		Link	Clamp, activate, bind, project		Move or Stay Put	Attach

Technical Terms			Biological Terms			
Functional Basis			Eng.-to-Bio. Thesaurus	Biologically Meaningful Keywords	Correlation Matrix NIST-BT	
Class	*Secondary*	*Tertiary*	*Function Correspondents*	*Keywords (% of colloc., # of matches)*	*Group*	*Sub-Group*
Connect	Mix		Blend, contract, exchange, fragment	Fragment (%26, #127), Exchange (%10, #220), Cleave (%9, #80), Bind (%6, #483), Break down (%4, #125), Contract (%1, #226), Cross over (%65, #34), Degrade (%8, #36)	Make	Chemically assemble
Control magnitude	Actuate		Activate, induce, trigger	Change shape (%11, #71), Bind (%9, #483), Stick (%6, #53), Activate (%5, #256), Change structure (%4, #53), Regulate (%3, #401), Absorb (%2, #172), Adapt (%7, #286), Evolve (%4, #424), Excite (%1, #69)		
	Regulate		Electrophoresis, gate, organogenesis, respire, sustain, preserve, remain, stabilize, maintain, regulate	Kill (%10, #102), Protect (%9, #161),		
		Increase	Hyperpolarize, pinocytosis, grow, expand, multiply, replicate	Relax (%29, #42), Stimulate (%18, #289), Activate (%14, #256), Contact (%10, #226), Project (%10, #50), Grow (%7, #786), Molt (%7, #44), Develop (%3, #843), Fold (%3, #74)	Maintain Physical Integrity	Regulate Physiological Processes
		Decrease	Compress, coil, divide, fold, shorten, wrap	Hyperpolarize (%21, #29), Oppose (%20, #15), Constrict (%8, #39), Stimulate (%3, #289), Inhibit (%3, #190), Narrow (%2, #47), Bind (%1, #483), Bulge (%6, #17)		

Technical Terms			Biological Terms			
Functional Basis			Eng.-to-Bio. Thesaurus	Biologically Meaningful Keywords	Correlation Matrix NIST-BT	
Class	*Secondary*	*Tertiary*	*Function Correspondents*	*Keywords (% of colloc., # of matches)*	*Group*	*Sub-Group*
Control magnitude	Change		Pinocytosis, degrade, alter, bind, catalyse, contract, hydrolysis, twist, mutate, radiate, charged, slip, acclimatize, alternate, fluctuate	Evolve (%7, #424), Specialize (%6, #164) Adapt (%1, #286)	Modify	Modify physical state Modify chemical/ electrical state
		Increment		Relax (%29, #42), Stimulate (%18, #289), Activate (%14, #256), Contact (%10, #226), Project (%10, #50), Grow (%7, #786), Molt (%7, #44), Develop (%3, #843), Fold (%3, #74)		
		Decrement	Decarboxylation, constrict	Hyperpolarize (%21, #29), Oppose (%20, #15), Constrict (%8, #39), Stimulate (%3, #289), Inhibit (%3, #190), Narrow (%2, #47), Bind (%1, #483), Bulge (%6, #17)		
		Shape	Elongate, stretch, attach, spread	Coil (%10, #30), Enlarge (%3, #33), Contract (%2, #226),		
		Condition	Osmosis, constrict			Adapt/ Optimize

Technical Terms			Biological Terms			
Functional Basis			Eng.-to-Bio. Thesaurus	Biologically Meaningful Keywords	Correlation Matrix NIST-BT	
Class	*Secondary*	*Tertiary*	*Function Correspondents*	*Keywords (% of colloc., # of matches)*	*Group*	*Sub-Group*
Control magnitude	Stop		Clog, extinguish, halt, interphase, seal, suspend	Lyse (%9, #23), Cut (%5, #134), Inhibit (%2, #190), Activate (%2, #256), Bind (%1, #483)		
		Prevent	Constrain, obstruct	Cover (%17, #121), Bind (%14, #483), Destroy (%10, #68), Stimulate (%9, #289), Surround (%9, #209), Inhibit (%7, #190), Release (%7, #508)	Maintain Physical Integrity	Manage structural forces Prevent structural failure
		Inhibit	Cover, destroy, inhibit, repress, surround		Maintain Physical Integrity	Protect from Biotic factors Protect from Abiotic factors Manage structural forces
Convert	Convert		Polymerize, synthesize, burn, gluconeogenesis, metabolize, grow, transduction, fermentation, glycolysis, hydrolysis, respiration, ionize, decompose, degrade, develop, mutate, photosynthesize	Specialize (%48, #164), Cut (%26, #134), Recombine (%26, #135), Transduce (%23, #99), Degrade (%14, #36), Synthesize (%14, #310), Photosynthesize (%13, #205), Stimulate (%13, #289), Transcribe (%12, #347), Fuse (%12, #120), Contract (%11, #226), Divide (%10, #277), Decompose (%10, #31), Break down (%9, #125), Activate (%8, #256), Mutate (%7, #299), Reproduce (%6, #537), Transpire (%37, #27), Coil (%10, #30)	Make Process information	Generate/ convert energy Encode/ Decode

Technical Terms			Biological Terms			
Functional Basis			Eng.-to-Bio. Thesaurus	Biologically Meaningful Keywords	Correlation Matrix NIST-BT	
Class	*Secondary*	*Tertiary*	*Function Correspondents*	*Keywords (% of colloc., # of matches)*	*Group*	*Sub-Group*
Provision	Store		Conserve, hold, convert, deposit, photosynthesize	Concentrate (%16, #58), Convert (%12, #146), Photosynthesize (%12, #205), Deposit (%10, #49), Dissolve (%7, #69)	Get, store, or distribute resources	Store
		Contain	Absorb	Enclose (%46, #78), Swell (%17, #35), Surround (%15, #209), Extend (%5, #95), Grow (%5, #786), Develop (%4, #843)		
		Collect	Absorb, catch, break-down, concentrate, Digest., reduce	Digest (%30, #267), Break down (%18, #125), Convert (%10, #146), Reduce (%9, #312), Feed (%9, #183), Cleave (%10, #80)	Get, store, or distribute resources	Capture, absorb, or filter
	Supply		Feed, lactate	Nurture (%17, #6), Break down (%5, #125), Convert (%4, #146), Degrade (%3, #36)	Maintain Community	Provide Ecosystem services
Signal	Sense			Receptor (%34, #503), Receive (%20, #172), Be stimulated (%14, #217), Bind (%14, #483), Curl (%50, #2), Protrude (%12, #17), Encounter (%4, #55)	Process information	Sense signal/ environment cues
		Detect	Detect, locate, see, small			
		Measure	Observe, monitor, gauge, watch	Emit (%48, #33), Recognize (%18, #203) Isolate (%9, #137)	Process information	Navigate Sense signal/ environment cues

Technical Terms			Biological Terms			
Functional Basis			**Eng.-to-Bio. Thesaurus**	**Biologically Meaningful Keywords**	**Correlation Matrix NIST-BT**	
Class	*Secondary*	*Tertiary*	*Function Correspondents*	*Keywords (% of colloc., # of matches)*	*Group*	*Sub-Group*
Signal	Indicate		Fluoresce, communicate, react, mark	Signal (%3, #399), Communicate (%4, #109)		
		Track				
		Display		Convey (%20, #10), Behave (%12, #374), Change shape (%8, #71), Signal (%8, #399), Bind (%5, #483), Stick out (%43, #7), Unwind (%17, #12), Denature (%14, #36), Change structure (%6, #53), Break down (%1, #125), Convert (%2, #146)	Process information	Send signal
	Process		Learn		Process information	Process signals Compute
Support			Develop, wrap	Anchor (%21, #24), Connect (%20, #167), Wrap (%13, #15), Divide (%3, #277), Bind (%2, #483), Develop (%2, #843)		
	Stabilize		Homeostasis, cling, hold, bind, connect			
	Secure		Surround, envelope			
	Position					

10 Index

H. Hashemi Farzaneh and U. Lindemann, *A Practical Guide to Bio-inspired Design*, https://doi.org/10.1007/978-3-662-57684-7

MIX
Papier aus verantwortungsvollen Quellen
Paper from responsible sources
FSC® C105338

If you have any concerns about our products,
you can contact us on
ProductSafety@springernature.com

In case Publisher is established outside the EU,
the EU authorized representative is:
Springer Nature Customer Service Center GmbH
Europaplatz 3, 69115 Heidelberg, Germany

Printed by Libri Plureos GmbH
in Hamburg, Germany